Cloud Computing

Data-Intensive Computing
and Scheduling

CHAPMAN & HALL/CRC
Numerical Analysis and Scientific Computing

Aims and scope:

Scientific computing and numerical analysis provide invaluable tools for the sciences and engineering. This series aims to capture new developments and summarize state-of-the-art methods over the whole spectrum of these fields. It will include a broad range of textbooks, monographs, and handbooks. Volumes in theory, including discretisation techniques, numerical algorithms, multiscale techniques, parallel and distributed algorithms, as well as applications of these methods in multi-disciplinary fields, are welcome. The inclusion of concrete real-world examples is highly encouraged. This series is meant to appeal to students and researchers in mathematics, engineering, and computational science.

Proposals for the series should be submitted to one of the series editors above or directly to:
CRC Press, Taylor & Francis Group
4th, Floor, Albert House
1-4 Singer Street
London EC2A 4BQ
UK

Published Titles

Classical and Modern Numerical Analysis: Theory, Methods and Practice
Azmy S. Ackleh, Edward James Allen, Ralph Baker Kearfott, and Padmanabhan Seshaiyer

Cloud Computing: Data-Intensive Computing and Scheduling
Frédéric Magoulès, Jie Pan, and Fei Teng

Computational Fluid Dynamics
Frédéric Magoulès

A Concise Introduction to Image Processing using C++
Meiqing Wang and Choi-Hong Lai

Decomposition Methods for Differential Equations: Theory and Applications
Juergen Geiser

Desktop Grid Computing
Christophe Cérin and Gilles Fedak

Discrete Variational Derivative Method: A Structure-Preserving Numerical Method for Partial Differential Equations
Daisuke Furihata and Takayasu Matsuo

Grid Resource Management: Toward Virtual and Services Compliant Grid Computing
Frédéric Magoulès, Thi-Mai-Huong Nguyen, and Lei Yu

Fundamentals of Grid Computing: Theory, Algorithms and Technologies
Frédéric Magoulès

Handbook of Sinc Numerical Methods
Frank Stenger

Introduction to Grid Computing
Frédéric Magoulès, Jie Pan, Kiat-An Tan, and Abhinit Kumar

Iterative Splitting Methods for Differential Equations
Juergen Geiser

Mathematical Objects in C++: Computational Tools in a Unified Object-Oriented Approach
Yair Shapira

Numerical Linear Approximation in C
Nabih N. Abdelmalek and William A. Malek

Numerical Techniques for Direct and Large-Eddy Simulations
Xi Jiang and Choi-Hong Lai

Parallel Algorithms
Henri Casanova, Arnaud Legrand, and Yves Robert

Parallel Iterative Algorithms: From Sequential to Grid Computing
Jacques M. Bahi, Sylvain Contassot-Vivier, and Raphael Couturier

Particle Swarm Optimisation: Classical and Quantum Perspectives
Jun Sun, Choi-Hong Lai, and Xiao-Jun Wu

Cloud Computing

Data-Intensive Computing and Scheduling

Frédéric Magoulès, Jie Pan,

and Fei Teng

SILKAN

CRC Press
Taylor & Francis Group
Boca Raton London New York

CRC Press is an imprint of the
Taylor & Francis Group, an **informa** business

A CHAPMAN & HALL BOOK

CRC Press
Taylor & Francis Group
6000 Broken Sound Parkway NW, Suite 300
Boca Raton, FL 33487-2742

© 2013 by Taylor & Francis Group, LLC
CRC Press is an imprint of Taylor & Francis Group, an Informa business

---

Library of Congress Cataloging-in-Publication Data

Magoulès, F. (Frédéric)
    Cloud-computing : data-intensive computing and scheduling / Frederic Magoules, Jie Pan, Fei Teng.
    p. cm. -- (Chapman & Hall/CRC numerical analysis and scientific computing series)
    Includes bibliographical references and index.
    ISBN 978-1-4665-0782-1 (hardback)
    1. Cloud computing. 2. Parallel programs (Computer programs) 3. Computer scheduling. I. Pan, Jie, 1979- II. Teng, Fei. III. Title.

QA76.585.M34 2012
004.6782--dc23
                                        2012015478

Visit the Taylor & Francis Web site at
http://www.taylorandfrancis.com

and the CRC Press Web site at
http://www.crcpress.com

Contents

List of Figures

List of Tables

Foreword

For a lot of people (even scientists and engineers) cloud computing is still a strange new paradigm. Some of them are even convinced that it is a new word for grid computing!

In fact, cloud computing is a new key discipline of high-performance computing (HPC) sophisticated IT technologies in order to treat some of the major HPC challenges and enabling trusted technical computing solutions for 21st century customers. Cloud computing is a technology that uses the Internet and central remote servers to maintain data and applications. Cloud computing allows consumers and businesses to use applications without installation and access their personal files at any computer with Internet access. This technology allows for much more efficient computing by centralizing storage, memory, processing, and bandwidth. Cloud computing provides ICT resources in a dynamic and scalable manner over a network. According to the National Institute of Standards and Technology, the five essential characteristics of the cloud are the following: on-demand self-service, broad network access, resource pooling, rapid elasticity, and measured services.

The applications of cloud computing are huge and impact nearly all sectors. For example, some automotive firms are convinced that the future lies in cloud computing: in some recent vehicles, the software used in the vehicle online applications system is not stored in the vehicle but runs off the firm backend. The advantage of this is that the applications can be continually updated in the cloud and new applications released to the automotive firm's customers without the need of visiting a workshop!

This time these tremendous civil applications are not derived from military applications such as the Internet or GPS. Cloud computing is only beginning to be used in Military Intelligence Fusion (MIF). The advantages of cloud computing are already applied to MIF in the sharing of data and application. Other known uses of cloud computing in the U.S. military sector are the following: the Rapid Access Computing Environment (RACE) used by the U.S. Department of Defense (DoD) and the U.S. military owned private cloud computing to Afghanistan. Nevertheless, cloud computing technology has not yet been effectively exploited in military embedded applications because of performance and correctness constraints. Some of these strategic embedded military applications are as follows: critical command and control systems, in-field data and information analysis, image data processing onboard missiles, UAVs, ... The main challenges of cloud computing are the need for wide bandwidth, concern for security, concern for malfunctions of the cloud, legal and political issues and concern for the rights of users. I am personally convinced that cloud computing will be relevant for new-generation high safety level critical systems (for both civil and military applications) because of the following important advantages: unlimited IT infrastructure flexibility and increased mission flexibility.

A strategic application of cloud computing is the treatment of the data wave. How to manage and derive continual value from your data? In terms of "Big Data", biology will pass physics within two years. For example, next-generation sequencers produce between six and eight TeraBytes (TB) of data per day! The challenges in next-generation genome centers are managing and processing oceans of data (TBs per day from multiple instruments, fast access and processing) and providing more services at a lower cost per genome.

The book by Frédéric Magoulès, Jie Pan, and Fei Teng is unique; it is the first book treating two key aspects of cloud computing: resource scheduling and allocation and "Big Data" treatment. For Big Data, the MapReduce model (parallel data flow system) is explained in detail and two implementation frameworks of MapReduce, Hadoop and Gridgain, are presented. These technologies enable us to "focus" the data wave in order to go to data analytics, visualization, and archiving. The book clearly highlights and presents these tools as a framework allowing distributed processing of large data sets across clusters of computers using a simple programming model. It also shows how they are designed to scale up from single servers to thousands of machines, each offering local computation and storage.

This book is addressing students specialized in high performance computing but also scientists and engineers from numerous sectors having to deal with Big Data: bioscience (pharmacological trials), financial and insurance services (automated and algorithm trading, fraud detection), science and research (large scale experiments as for example the large hadron collider, satellites feeds, medical imaging, . . .), educational research, legacy (sales data, accounting data and customer data), retailers (customer buying analysis, inventory management), government and military agencies (signal analysis, trend analysis).

This book will also allow a deeper understanding of the technical characteristics of the major best-in-class cloud computing providers (public, private, government) and associated services such as Infrastructure as a Service (IaaS) and Software as a Service (SaaS). It also highlights tutorial approaches to the main cloud characteristics such as global reach, ease of provisioning, business agility, deployability, and manageability.

Jacques Duysens
General Manager - Chief Operating Officer
SILKAN

Preface

Cloud computing has emerged as a hot topic of research, and several books have been published on this topic. This present book concentrates on data-intensive computing and scheduling for cloud computing, with a particular focus on new development of classical techniques, and recent methods and innovative algorithms appearing in this field. This volume presents in nine chapters a selection of some concepts, models, methods, algorithms, and software used in cloud computing including resource management, MapReduce, multi-dimensional data analysis, multiple group-by query, and real-time scheduling.

Chapter 1 begins with a general introduction of cloud computing. Although there is disagreement over what cloud computing is, Chapter 1 tries to refine some representations and gives an unbiased and general definition. This definition is not just an overall concept, but describes system architecture, deployment models, and essential features. Cloud computing is still an evolving paradigm, and it integrates many existing technologies. The brief evolution history of cloud computing described in Chapter 1 helps us clarify the conditions, opportunities, and challenges existing in cloud development. Functionally speaking, cloud computing is a service provision model, in which software, platform, infrastructure, data, and hardware can be directly delivered as a service to end customers. Chapter 1 presents the service characteristics from technical, qualitative, and economic points of view. After analyzing existing commercial products and research projects, several challenges in terms of middleware, programming model, resource management, and business model are highlighted.

In Chapter 2, the cloud service scheduling hierarchy is presented in detail together with scheduling problems. The scheduling problems can be split into a user level and a system level. The former focuses on the issue of resource provision between providers and customers, which is solved by economic models. The latter refers to meta-task execution, a sub-optimal solution of which is given by heuristics to speed up the process of finding a good enough answer. For commercial purposes, cloud services heavily emphasize time guarantee. The ability to satisfy timing constraints of such real-time applications plays a significant role in the cloud environment. Chapter 2 examines and introduces some particular scheduling algorithms for real-time tasks, that is, priority-based strategies. Their implementations are discussed in depth and analyzed to match real-time constraints.

Cloud computing can cut IT costs and at the same time herald in a new era of agility in IT operations. A fundamental element is the concept of a datacenter, in which IT solutions are considered as services and are as easily purchased as other consumption models. Therefore, resource provision takes on market dealing behaviors, not just match-making scheduling between tasks and machines. The market mechanism is an effective method to control electronic resources. In Chapter 3, resource competition among cloud customers and reasonable allocation to keep market equilibrium, are analyzed. Specifically, Chapter 3 presents a game theoretical auction to solve the resource allocation problem in clouds, and proposes original practicable algorithms for user bidding and auctioneer pricing. Such algorithms support financially smart customers with an effective forecasting method and help an auctioneer decide on an equilibrium resource price, so that they can potentially solve resource allocation problems in cloud computing.

An another important aspect of cloud computing is multi-dimensional data analysis applications, since enterprises generate massive amounts of data every day. To analyze this data, the raw data is extracted, transformed, cleaned, stored under multi-dimensional data models, and finally queried by the user. Recently, new technologies have been adopted in multi-dimensional data analysis applications. These new technologies include in-memory query processing, search engine technologies, and enhanced hardware. In Chapter 4, the features of multi-dimensional data analysis queries and three distributed system architectures including shared-memory, shared-disk, and shared-nothing are described. A survey of existing research on accelerating data analytical query processing is also provided, including pre-computing, data indexing, and data partitioning. Pre-computing is an approach to bartering storage space for computing time. The aggregates of all possible dimension combinations are calculated and stored to rapidly answer the forthcoming queries. Data indexing technologies can appear in several forms such as B-tree/B^+-tree index, projection index, Bitmap index, Bit-Sliced index, join index, inverted index, together with a special type of index used in distributed architecture. Classical data partitioning technology consists of horizontal partitioning, or vertical partitioning. Chapter 4 presents the application of partitioning methods on a multi-dimensional dataset, followed by the parallelization of query processing. A special emphasis on the parallelization of various operators, including scan, merge, split, selection, update, sorting, aggregation, duplicate removal, and join is presented in order to ensure multi-dimensional data analysis in a cloud data center.

As mentionned previously, along with the development of hardware and software, more and more data is generated at a rate much faster than ever. Although data storage is inexpensive, and the issues of storing large volumes of data can be solved, processing large volumes of data is becoming a challenge for data analysis software. The feasible approach to handling large-scale data processing is to *divide and conquer*. Solutions based on a parallel model, for instance the parallel database composed of shared-nothing distributed architectures, can be considered. Relations are partitioned into pieces of data, and the computations of one relational algebra operator proceeded in parallel on each piece of data. The traditional parallel attempts in data intensive processing, like parallel database, were suitable when data scale was moderate because parallel databases do not scale well. MapReduce is a new parallel programming model, which turns a new page in data parallelism history. The MapReduce model is a parallel data flow system that works through data partitioning across machines, each machine independently running the single-node logic. Chapter 5 focuses on the MapReduce model, and its extended model, MapCombineReduce. Two implementation frameworks of MapReduce, Hadoop and GridGain are presented as an example. Job scheduling issues in MapReduce are then analyzed, followed by the distributed data storage underlying MapReduce, including distributed file systems and an efficient enhanced storage system based on a cache mechanism. Finally, Chapter 5 discusses transactional data management and analytical data management in the cloud processed with MapReduce. As shared-nothing parallel databases and MapReduce systems use similar hardware, Chapter 5 focuses on comparing them and by presenting the related work on a hybrid solution combining these two into one system.

In Chapter 6, in order to manipulate large-scale multi-dimensional data, MapReduce-based multi-dimensional data aggregation is presented, followed by the introduction of multiple group-by query. GridGain is chosen over Hadoop, as the MapReduce supporting framework because of its low latency. A detailed workflow analysis of the GridGain MapReduce procedure is presented. Two implementations of multiple group-by query based on MapReduce, initial and optimized implementations are illustrated. The initial implementation of the multiple group-by query is based on a direct realization, which implements the filtering phase within mappers and the aggregating phase within the reducer. In the optimized implementation of the multiple group-by query, a combiner as a pre-aggregator, which does the aggregation (pre-aggregation) on a local computing node level before starting the reducer, is adopted. With such a pre-aggregator, the amount of intermediate data transferred over the network is significantly reduced. As GridGain does not support a combiner component, the combiner through merging two successive GridGain's MapReduces is constructed. Chapter 6 presents experiments run on a public French academic platform named Grid'5000, which demonstrates that the optimized version has better speed and better scalability. At the end of Chapter 6, a formal estimation of execution time is given for both implementations; estimations which could also serve as a valuable reference for other MapReduced applications.

In a distributed shared-nothing architecture, like the MapReduce system, there are two approaches to optimize query processing. The first one is to choose an optimal job scheduling policy in order to complete the calculation within a minimum time. Load balancing, data skew, and straggler node are some issues involved in job scheduling. The second approach focuses on the optimization of individual jobs constituting the parallel query processing. Individual job optimization needs to consider the characteristics of involved computations, including the low-level optimization of detailed operations. Chapter 7 discusses the optimization work for accelerating individual jobs during the parallel processing procedure of the multiple group-by query. Then, the speed-up performance of implementations over horizontally partitioned data and that of vertically partitioned data during this procedure are presented. An estimation model for the query processing execution time, and specifically for estimating the values of various parameters for data horizontal partitioning-based query processing is then presented. Finally, Chapter 7 introduces a new compressed data structure, which works with vertical partition in order to support distinct-value-wise job scheduling.

Since MapReduce has been beneficial to a wide spectrum of data-intensive applications such as search indexing, mining social networks, recommendation services, and advertising backends, accurate time guarantee turns out to be more important for better QoS than ever. Therefore, the problem of scheduling real-time tasks on a MapReduce cluster will be investigated in Chapter 8. A MapReduce scheduling algorithm combining the particular characteristics of MapReduce is thus introduced, followed by some tuning leading to enhanced scheduling efficiency. Finally, Chapter 8 proposes an original method to indicate the reliability of a schedulability test. From the aspect of system, a test with high reliability can guarantee high system utilization.

Cloud computing implies that computing is not only operated on local computers, but on centralized facilities by third-party computing and storage utilities. Cloud solutions seem to state master keys for the IT enterprises that suffer from budget concerns and economic woes, and a number of industry projects have been started to create a global, multi-data center, open-source cloud computing testbed for industry, research, and education. Encouraging opportunities also brings out corresponding challenges. Cloud computing is easily confused with several existing technologies including grid computing, utility computing, web service, and virtualization. Scheduling problems in cloud computing are worth reconsidering by researchers and engineers. In this book, the resource allocation problem in terms of economic aspects to meet business requirements is addressed, together with the real-time schedulability constraint to provide the cloud data center with technical supports. In order to utilize cloud computing to serve as the infrastructure of multi-dimensional data analysis applications, the combination of traditional parallel database optimization mechanisms and cloud computing is expected. In this book, classical or original methods of cloud computing to satisfy commercial software requirements are presented. Chapter 9 draws a brief summary of this book and raises many interesting questions and issues that deserve further research, including (i) choosing suitable serialization or de-serialization algorithms to deal with mapper objects and intermediate results; (ii) extending calculations to a larger computing scale; (iii) utilizing cloud computing to process large datasets; (iv) enriching business models for cloud providers; (v) expanding schedulability bound to more complicated systems; (vi) improving reliability of on-line schedulability tests for cloud data centers. The performance issues addressed in this book represent an important aspect in cloud computing, and could provide a useful reference for people who want to study and utilize MapReduce and cloud computing platforms.

The various technologies presented in this book demonstrate the wide aspects of interest in cloud computing, and the many possibilities and venues that exist in the research in this area. This interest is only going to further evolve, and many exciting developments are still awaiting us.

Frédéric Magoulès
Ecole Centrale Paris, France

Jie Pan
Klee Group, France

Fei Teng
Southwest Jiaotong University, China

Warranty

Every effort has been made to make this book as complete and as accurate as possible, but no warranty of fitness is implied. The information is provided on an as-is basis. The authors, editor, and publisher shall have neither liability nor responsibility to any person or entity with respect to any loss or damages arising from the information contained in this book or from the use of the code published in it.

Chapter 1

Overview of cloud computing

1.1 Introduction

This chapter begins with the general introduction of cloud computing, followed by the retrospect of cloud evolution history and comparison with several related technologies. Through analyzing system architecture, deployment model and service type, the characteristics of cloud computing are concluded from technical, functional and economic aspects. After that, current efforts both from commercial and research perspectives are presented in order to capture challenges and opportunities in this domain.

1.1.1 Cloud definitions

Since 2007, the term Cloud has become one of the most common buzz words in the IT industry. Lots of researchers try to define cloud computing from different application aspects, but there is not a consensus definition of it. Among the many definitions, we choose three widely quoted as follows

> **I. Foster** [Foster et al., 2008]: "A large-scale distributed computing paradigm that is driven by economies of scale, in which a pool of abstracted virtualized, dynamically-scalable, managed computing power, storage, platforms, and services are delivered on demand to external customers over the Internet."

As an academic representative, Foster focuses on several specific features that differ from other distributed computing paradigms. Cloud computing, in which computing entities are virtualized and delivered as services, is massively scalable. These services are dynamically configured and driven by economies of scale.

> **Gartner** [Plummer et al., 2008]: "A style of computing where scalable and elastic IT capabilities are provided as a service to multiple external customers using Internet technologies."

Gartner, being an IT consulting company, examines qualities of cloud computing mostly from the point of view of industry. Functional characteristics are emphasized in this definition, such as whether cloud computing is scalable, elastic, service offering, and Internet based.

NIST [Mell and Grance, 2010]: "Cloud computing is a model for enabling convenient, on-demand network access to a shared pool of con-
figurable computing resources (e.g., networks, servers, storage, applications, and services) that can be rapidly provisioned and released with
minimal management effort or service provider interaction."

Compared with the above two definitions, U.S. National Institute of Standards and
Technology provides a relatively more objective and specific definition, which not
only defines cloud concept overall, but also specifies essential characteristics of cloud
computing, delivery and deployment models.

1.1.2   System architecture

Clouds are usually referred to as a large pool of computing and/or storage resources, which can be accessed via standard protocols with an abstract interface [Fos-
ter et al., 2008]. There is four-layer architecture for cloud computing as shown in
Figure 1.1. The fabric layer contains the raw hardware level resources, such as com-

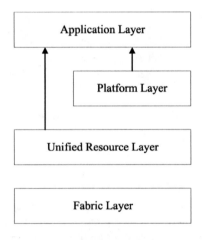

FIGURE 1.1: System architecture.

pute resources, storage resources, and network resources. The unified resource layer
contains resources that have been virtualized so that they can be exposed to upper
layer and end users as integrated resources. The platform layer adds a collection of
specialized tools, middleware, and services on top of the unified resources to provide a development and/or deployment platform. The application layer contains the
applications that would run in the clouds.

1.1.3 Deployment models

Clouds can be deployed in different fashions, depending on the usage scopes. There are four primary cloud deployment models.

Public cloud. The public cloud is the standard cloud computing paradigm, in which a service provider makes resources, such as applications and storage, available to the general public over the Internet. Service providers charge on a fine-grained utility computing basis. Examples of public clouds include Amazon Elastic Compute Cloud (EC2), IBM's Blue Cloud, Sun Cloud, Google AppEngine and Windows Azure Services Platform.

Private cloud. A private cloud looks more like a marketing concept than the traditional mainstream sense. It describes a proprietary computing architecture that provides services to a limited number of people on internal networks. Organizations needing accurate control over their data will prefer a private cloud, so they can get all the scalability, metering, and agility benefits of a public cloud without ceding control, security, and recurring costs to a service provider. Both eBay and HP CloudStart yield private cloud deployments.

Hybrid cloud. A hybrid cloud uses a combination of public cloud, private cloud and even local infrastructures, which is typical for most IT enterprises [Keith and Burkhard, 2010]. Hybrid strategy is proper placement of workloads depending upon cost and operational and compliance factors. Major vendors including HP, IBM, Oracle and VMware create appropriate plans to leverage a mixed environment, with the aim of delivering services to the business. Users can deploy an application hosted on a hybrid infrastructure, in which some nodes are running on real physical hardware and others on cloud server instances.

Community cloud. A community cloud overlaps with Grids to some extent. It refers to several organizations in a private community sharing the cloud infrastructure. The organizations usually have similar concerns about mission, security requirements, policy, and compliance considerations. A community cloud can be further aggregated by a public cloud to build up a cross-boundary structure.

1.1.4 Cloud characteristics

As a general resource provisioning model, cloud computing integrates a number of existing technologies that have been applied in grid computing [Magoulès et al., 2009], [Magoulès, 2009], utility computing, service oriented architectures, internet of things, outsourcing, etc. That is the reason why the cloud is mistaken for "the same old stuff with a new label." In this section, we distinguish between the technical, qualitative, and economic aspects of cloud computing.

Technical aspects.   Technical characteristics are the foundation that ensures other functional and economic requirements. Not all technology is (absolutely) new, but might be enhanced to realize a specific feature, directly or as a pre-condition.

- **Virtualization.** Virtualization is an essential characteristic of cloud computing. Virtualization in clouds refers to a multi-layer hardware platform, operating system, storage device, network resources, etc.

  The first prominent feature of virtualization is the ability to hide the technical complexity from users, so it can improve the independence of cloud services. The second is that, physical resources can be efficiently configured and utilized, considering that multiple applications are run on the same machine. Third, quick recovery and fault tolerance are permitted, because the virtual environment can be easily backed up and migrated with no interruption in service [Cafaro and Aloisio, 2010].

- **Multi-tenancy.** Multi-tenancy is a highly requisite issue in clouds, which allows sharing of resources and costs across multiple users.

  Multi-tenancy brings resource providers many benefits, for example, centralization of infrastructure in locations with lower costs, and improvement of utilization and efficiency with high peak-load capacity. Tenancy information, which is stored in separate databases but altered concurrently, should be well maintained for isolated tenants. Otherwise, some problems such as data protection will arise.

- **Security.** Security is one of the largest concerns for adoption of cloud computing. There is no reason to doubt the importance of security in any system dealing with sensitive and private data. In order to gain the potential clients, providers must supply the certificate of security. For example, data should be fully segregated from one to another, and an efficient replication and recovery mechanism should be prepared if a disaster occurs. Besides that, a contractual commitment is desired or ensured for investigative support.

  In terms of complexity, on the one hand, the complexity of security is increased when data is distributed over a wider area or greater number of devices in multi-tenant systems which are shared by unrelated users. On the other hand, the complexity reduction is necessary, because "ease of use" ability can attract more potential clients.

- **Programming environment.** Programming environment is essential to exploit cloud features. It should be capable of addressing issues such as multiple administrative domains, large variations in resource heterogeneity, performance stability, exception handling in highly dynamic environments, etc.

  System interface adopts web services' APIs, which provide a standards-based framework for accessing and integrating with and among cloud services. A browser, applied as the interface, has attributes such as being intuitive, easy-to-use, standards-based, service-independent and multi-platform supported.

Through pre-defined APIs, users can access, configure and program cloud services.

Qualitative aspects. Qualitative characteristics refer to qualities or properties of cloud computing, rather than specific technological requirements. One qualitative feature can be realized in multiple ways depending on different providers.

- **Elasticity.** Elasticity means that the provision of services is elastic and adaptable, which allows the users to request the service near real-time without engineering for peak loads. The services are measured in fine-grain, so that the amount of offering can perfectly match a consumer's usage. Performance is monitored and consistent.

- **Availability.** Availability refers to a relevant capability that satisfies specific requirements of the outsourced services. In many use cases, QoS metrics such as response time and throughput must be guaranteed, so as to ensure advanced quality guarantees of cloud users.

- **Reliability.** Reliability represents the ability to ensure constant system operation without disruption. Using redundant sites, the possibility of losing data and code dramatically decreases so that cloud computing is suitable for business continuity and disaster recovery. Reliability is a particular QoS requirement, focusing on prevention of loss.

- **Agility.** Agility is a basic requirement for cloud computing. Cloud providers should be capable of on-line reaction to changes in resource demand and environmental conditions. At the same time, efforts are made by clients to re-provision an application from an in-house infrastructure to SaaS vendors. Agility requires both sides to provide self management capabilities.

Economic aspects. Economic features make cloud computing distinct, compared with other computing paradigms. In a commercial environment, service offerings are not limited to an exclusive technological perspective, but extend to a broader understanding of business needs.

- **Pay-as-you-go.** Pay-as-you-go is a common aproach to cloud computing, which means users pay according to the actual consumption of resource. Traditionally, users have to be equipped with all software and hardware infrastructure before computing starts, and maintain it during the computing process. Cloud computing reduces cost of infrastructure maintenance and acquisition, so it can help enterprises, especially small to medium sized, reduce time to market and get return on investment.

- **Operational expenditure.** Operational expenditure is greatly reduced and converted to operational expenditure [Böhm et al., 2010]. The infrastructure is typically provided by a third-party and does not need to be purchased for one-time or infrequent intensive computing tasks, so it is easier for the users to enter the computing world. Minimal or no IT skills are required for implementation. Pricing on a utility computing basis is fine-grained with usage-based options, so cloud providers should mask this pricing granularity with long-term, fixed price agreements considering the customer's convenience.

- **Energy-efficiency.** Energy-efficiency is due to the ability of clouds to reduce the consumption of unused resources. Computers are administrated centrally, so additional costs of energy consumption as well as carbon emission can be better controlled than in uncooperative cases. In addition, green IT issues are subject to both software stack and hardware level.

1.2   Cloud evolution

Although the idea of cloud computing is not new, it has rapidly become a new trend in the information and communication technology domain and has gained significant commercial success over past years. No one can deny that cloud computing will play a pivotal role in the next decade. Why has cloud computing not emerged before? This section looks back over the development history of cloud computing.

1.2.1   Getting ready for the cloud

Datacenter.   Even faster than Moore's law, the number of servers and datacenters has increased dramatically over the past few years. Datacenter has turned out to become the reincarnation of the mainframe concept. It is easier to build a large-scale commodity-computer datacenter than ever before, just gathering these building blocks together on a parking lot and connecting them to the Internet.

Internet.   Recently, network performance has been improving rapidly. Wired, wireless and 4th generation mobile communication make Internet available to most of the planet. Cities and towns are wired with hotspots. Transportation such as air, train or maritime is also equipped with satellite based wi-fi or undersea fiber-optic cable. People can connect to Internet, virtually anywhere and at anytime. The universal, high-speed, broadband Internet laid the foundation for the widespread applications of cloud computing.

Terminals.   The PC is no longer the central computing device, various electronic devices including MP3, SmartPhone, Tablet, Set-top box, PDA, notebook have become new terminals that meet the personal computing requirement. Besides, repeated data synchronization among different terminals is time-consuming, and frequent faults occur. In such cases, a solution that allows individuals to access personal data anywhere and from any device is needed.

1.2.2 Brief history

Along with the maturity of objective conditions (software, hardware), many existing technologies, results and ideas can be realized, updated, merged and further developed.

Amazon played a key role in the development of cloud computing by initially renting their datacenter to external customers for personal computing use. In 2006, they launched Amazon EC2 and S3 on a utility computing basis. After that, several major vendors released cloud solutions one after another, including Google, IBM, Sun, HP, Microsoft, Forces.com, Yahoo and so on. Since 2007, the number of trademarks covering cloud computing brands, goods, and services has increased at an almost exponential rate.

At the same time, cloud computing is also a much favored research topic. In 2007, Google, IBM, and a number of universities announced a research project, the Academic Cloud Computing Initiative (ACCI), aimed at addressing the challenges of large-scale distributed computing. Since 2008, several open source projects have gradually appeared. For example, Eucalyptus is the first API-compatible platform for deploying private clouds. OpenNebula deploys private and hybrid clouds and federates different modes of clouds.

In July 2010, SiteonMobile was announced by HP for emerging markets where people are more likely to access the Internet via mobile phones rather than computers. With more and more people owning smartphones, mobile cloud computing has become a potent trend. Several Mobile network operators such as Orange, Vodafone and Verizon have started to offer cloud computing services for companies.

In March 2011, the Open Networking Foundation which consists of 23 IT companies, was founded by Deutsche Telekom, Facebook ®, Google ®, Microsoft ®, Verizon ®, and Yahoo ®. This nonprofit organization supports a new cloud initiative called Software-Defined Networking. The initiative is meant to speed up innovation through simple software changes in telecommunications networks, wireless networks, datacenters and other networking areas.

A simple account of cloud development history is presented in Figure 1.2.

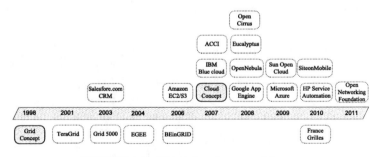

FIGURE 1.2: Cloud development history.

1.2.3   Comparison with related technologies

Cloud computing is a natural evolution of widespread adoption of virtualization, service-oriented architecture, autonomic and utility computing. It emerges as a new computing paradigm to provide reliable, customized, and quality of service guaranteeing dynamic computing environments for end-users, so it is easily confused with several similar computing paradigms such as utility computing, grid computing, and autonomic computing.

Utility computing.   Utility computing was initialized in the 1960s. John McCarthy coined the term computer utility in a speech given to celebrate MIT's centennial "If computers of the kind I have advocated become the computers of the future, then computing may someday be organized as a public utility just as the telephone system is a public utility. The computer utility could become the basis of a new and important industry." Generally, utility computing considers the computing and storage resources as a metered service like water, electricity, gas, and telephony utility. The customers can use the utility services immediately, whenever and wherever they need, without paying for the initial cost of the devices. This idea was very popular in the late 1960s, but faded by the mid-1970s as the devices and technologies of that time were simply not ready. Recently, the utility idea has resurfaced in new forms such as grid computing and cloud computing.

Utility computing is virtualized so that the amount of storage or computing power available is considerably larger than that of a single time-sharing computer. The back-end servers such as the computer cluster and supercomputer are used to realize the virtualization.

Since the late 90s, utility computing has resurfaced. HP launched the Utility Data Center to provide the IP billing-on-tap services. PolyServe Inc. offers a clustered file system based on commodity server and storage hardware that creates highly available utility computing environments for mission-critical applications and workload optimized solutions, specifically tuned for bulk storage, high-performance computing, vertical industries such as financial services, seismic processing, and content serving. Thanks to these utilities, including database and file service, customers can independently add servers or storage as needed.

Grid computing.   Grid computing emerged in the mid 90's. Ian Foster integrated distributed computing, object-oriented programming and web services to "coin the grid computing infrastructure." "A Grid is a type of parallel and distributed system that enables the sharing, selection, and aggregation of geographically distributed autonomous resources dynamically at runtime depending on their availability, capability, performance, cost, and users' quality-of-service requirements." [Foster et al., 2001] The definition explains that a grid is actually a cluster of networked, loosely coupled computers which works as a super and virtual mainframe to perform thousands of tasks. It can also divide the huge application job into several subjobs and make each run on large-scale machines.

Generally speaking, grid computing goes through three different generations [Roure et al., 2004]. The first generation was marked by the early metacomputing environment, such as FAFNER and I-WAY. The second generation was represented by the development of core grid technologies, grid resource management (e.g., GLOBUS, LEGION); resource brokers and schedulers (e.g., CONDOR, PBS) and grid portals (e.g., GRID SPHERE). The third generation saw the convergence between grid computing and web services technologies (e.g., WSRF, OGSI). It moves to a more service oriented approach that exposes the grid protocols using web service standards.

Autonomic computing. Autonomic computing was first proposed by IBM in 2001, that is "autonomic computing performs tasks that IT professionals choose to delegate to the technology according to policies. Adaptable policy rather than hard coded procedure determines the types of decisions and actions that autonomic capabilities perform." [Parashar and Hariri, 2005] Considering the sharply increasing number of devices, the heterogeneous and distributed computing systems are more and more difficult to anticipate, design and maintain. The complexity of management is becoming the limiting factor of future development. Autonomic computing focuses on the self-management ability of the computer system. It overcomes the rapidly growing complexity of computing systems management and reduces the barriers that complexity poses to further growth.

In the area of multi-agent systems, several self-regulating frameworks are proposed, but most of these architectures are centralized, which mainly reduces management costs and seldom considers enabling complex software systems and providing innovative services. IBM defined the self-managing system which can automatically process configuration of the components (Self-Configuration), automatic monitoring and control of resources to ensure the optimal (Self-Healing), monitor and optimize the resources (Self-Optimization) and proactive identification and protection from arbitrary attacks (Self-Protection), only with the input information of policies defined by humans [Kephart and Chess, 2003]. In other words, the autonomic system uses high-level rules to check and optimize its status and automatically adapt itself to changing conditions.

According to the above introductions to the three computing paradigms, we can summarize the relationship among them. Utility computing is concerned with whether the packing computing resources can be used as a metered service on the basis of the user's needs. It is indifferent to the organization of the resources, both in the centralized and distributed system. Grid computing is conceptually similar to the canonical Foster definition of cloud computing, but does not consider the economic entities. Autonomic computing stresses the self management of computer systems, which is only one feature of cloud computing. All in all, having grid technologies, autonomic characteristics, and utility bills, cloud computing can be seen as a natural next step from the grid-utility model.

1.3   Cloud services

Since a cloud is an underlying delivery mechanism, computing ability can be provisioned as services, basically on three levels: software, platform, and infrastructure [Armbrust et al., 2009].

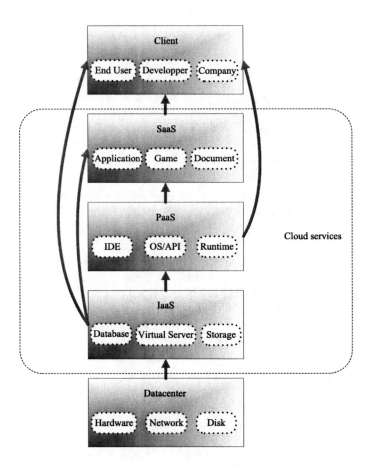

FIGURE 1.3: Cloud services.

Software as a service.   Software as a Service (SaaS) is a software delivery model in which applications are accessed by users using a simple interface such as a web browser over the Internet. The users are not concerned with the underlying cloud infrastructure including network, servers, operating systems, storage, platform, etc.

This model also eliminates the need to install and run the application on the local computers. The term SaaS has been popularized by Salesforce.com, which distributes business software on a subscription basis, rather than on a traditional on-premise basis. One of the best well known is the solution for its Customer Relationship Management (CRM). Now SaaS has become a common delivery model for most business applications, including accounting, collaboration and management. Applications such as social media, office software, and online games enrich the family of SaaS-based services, for instance, web Mail, Google Docs, Microsoft online, NetSuit, MMOG Games, Facebook, etc.

Platform as a service. Platform as a Service (PaaS) offers a high-level integrated environment to build, test, deploy and host customer-created or acquired applications. Generally, developers accept some restrictions on the type of software they can write in exchange for built-in application scalability. Customers of PaaS do not need to manage the underlying infrastructure as SaaS users, but need to have control over the deployed applications and their hosting environment configurations.

PaaS offerings mainly aim at facilitating application development and related management issues. Some are intended to provide a generalized development environment, and some only provide hosting-level services such as security and on-demand scalability. Typical examples of PaaS are Google App Engine, Windows Azure, Engine Yard, Force.com, Heroku, MTurk, etc.

Infrastructure as a service. Infrastructure as a Service (IaaS) provisions processing, storage, networks, and other fundamental computing resources to users. IaaS users can deploy and run arbitrary applications, software, and operating systems on the infrastructure that can scale up and down dynamically based on resource needs.

Computing service allows users to rent a provider's virtual machines, or even an entire datacenter. The user sends programs and related data, while the vendor's computer does the computation processing and returns the result. The infrastructure is virtualized, flexible, scalable, and manageable to meet user needs. Examples of IaaS include Amazon EC2, VPC, IBM Blue Cloud, Eucalyptus, FlexiScale, Joyent, Rackspace Cloud, etc.

Data service concerns the access of users to remote data in various formats and from multiple sources. This remote data can be operated just like on a local disk. Amazon S3, SimpleDB, SQS and Microsoft SQL are data service products.

Figure 1.3 shows the relationship among cloud users, cloud services, and cloud providers. Clients equipped with basic devices, Internet and web browsers can directly use software, platform, storage, and computing resources as pay-as-you-go services. If an Internet protocol connection is established, cloud services can be shared within any one of the service layers. For example, PaaS can consume IaaS offerings, and meanwhile, deliver platform supporting services to SaaS. At the bottom, the datacenter consists of computer hardware and software products that are specifically designed for the delivery of cloud services, including cloud-specific operating systems, multi-core processors, networks, disks, etc.

1.4   Cloud projects

We finish the state-of-the-art efforts made by both commercial and academic circles. Major vendors have invested in forthright progress in the area of global cloud promotion, while comparatively, research organizations based on their funding principles and interest, contribute to cloud technologies in an indirect way.

1.4.1   Commercial products

In the last few years, middleware and platforms which involve multiple level services in heterogeneous, distributed systems, have emerged. Commercial cloud solutions have augmented dramatically and promote organizational shift from company-owned assets to per-use service-based models. The best known cloud projects are Amazon Web Service, IBM SmartCloud, Eucalyptus, FlexiScale, Joyent, Azure, Engine Yard, Heroku, Force.com, RightScale, Netsuite, Google Apps, etc.

Amazon is the pioneer of cloud computing. In 2002, Amazon began to provide online computing services through the Internet. End users, not limited to developers can access these web services over HTTP, using Representational State Transfer and SOAP protocols. All services are billed on usage, but how usage is measured for billing varies from service to service [Yi et al., 2010]. Among them, the two most popular are Amazon EC2 and Amazon S3, which are typical representatives of IaaS. The former rents virtual machines for running local computing applications, and the latter offers online storage, which has the same infrastructure as Amazon.com uses.

Amazon EC2.   Amazon EC2 [Amazon EC2, 2011] allows users to create a virtual machine, named *instance*, through an Amazon Machine Image. An instance functions as a virtual private server that contains desired software and hardware. Roughly, instances are classified in six categories: standard, micro, high-memory, high-CPU, cluster-GPU, and cluster compute, each of which is subdivided into different configurations, such as memory, number of virtual cores, storage, platform, I/O performance and API. Besides that, EC2 supports security control of network access, instance monitoring, multi-location processing, etc.

Amazon S3.   Amazon S3 [Amazon S3, 2011] provides a highly durable storage infrastructure that can be used to store and retrieve data on the internet. This service is beneficial to developers by making computing more scalable. S3 stores data redundantly on multiple devices and supports version control to recover from both unintended user actions and application failures

Google App Engine.   Google App Engine [Google App Engine, 2011], released in 2008, is a platform for developing and hosting web applications in multiple servers and data centers. In terms of PaaS, GAP is written to be language dependent, and only supports Python and Java, so the runtime environment on GAP is limited. Compared to IaaS, GAP makes it easy to develop scalable applications, but can only run a limited range of applications designed for that infrastructure.

MapReduce.   MapReduce [Dean and Ghemawat, 2008] is the best known programming model introduced by Google that can support distributed computing on large clusters. It can carry out map and reduction operations process in parallel. The advantage of MapReduce is that it can efficiently handle significantly larger datasets than common servers and that it can quickly recover from partial failure of servers or storage during the operation. MapReduce is widely used both in industry and academic research. Google developed patented framework, while the Hadoop is open source with free license. Moreover, many projects like Twister, Greenplum, GridGain, Phoenix, Mars, CouchDB, Disco, Skynet, Qizmt, and Meguro implement the MapReduce programming model in different languages including C++, C#, Erlang, Java, Ocaml, Perl, Python, Ruby, etc.

Dryad.   The Dryad [Dryad, 2011] processing framework was developed by Microsoft as a declarative programming model on top of the computing and storage infrastructure. DryadLINQ aims at writing large-scale data parallel applications on large dataset clusters of computers.   DryadLINQ enables developers to use thousands of machines, each with multiple processors or cores, without knowing anything about concurrent programming. It supports automatic parallelization and serialization by translating LINQ programs into distributed Dryad computations.

Other common programming models include All-Paris, Mesh-up, Sector/Sphere and Mortar, etc.

1.4.2   Research projects

Besides initiatives by enterprises, a number of academic projects have been developped to address challenges including stable testbed, standardization, open source reference implementation, and more open source solutions. The most active projects in Europe and North America include XtreemOS, OpenNebula, FutureGrid, elasticLM, gCube, ManuCloud, RESERVOIR, SLA@SOI, Contrail, ECEE, NEON, VMware, Tycoon, DIET, BEinGRID, etc.

XtreemOS.   XtreemOS [Xtreemos, 2011] is an open source distributed operation system for grids. The project was initialized by INRIA in 2006, and the first stable release published in 2010.

XtreemOS strives to be a uniform computing platform by integrating heterogeneous infrastructures, from mobile devices to clusters. It provides three services including application execution management, data management, and virtual organization management.

Although XtreemOS was originally designed for grids, it can also be seen as an alternative for cloud computing, owing to the fact that it is relevant in the context of virtualized distributed computing infrastructure. Hence, its potential is to support cooperation and resource sharing over cloud federations.

OpenNebula.   OpenNebula [OpenNebula, 2011] is an open source project aiming at managing datacenters' virtual infrastructure to build IaaS clouds. It was established by Complutense University of Madrid in 2005, and released its first software in 2008.

It supports private cloud creation based on local virtual infrastructure in datacenters, and has the capabilities in management of user, virtual network, multi-tier services, and physical infrastructures. It also supports combination of the local resources and remote commercial clouds to build hybrid clouds, in which local computing capacity is supplemented by single or multiple clouds to better serve user access requests. In addition, it can be used as interfaces to turn local infrastructure into a public cloud.

FutureGrid.   FutureGrid [FutureGrid, 2011] is a test-bed for grid and cloud computing. It is a cooperative project launched in 2010 between Grid'5000 and TeraGrid.

FutureGrid builds the federation of multiple clouds with a large geographical distribution, and allows researchers to study the range from authentication, authorization, scheduling, virtualization, middleware design, interface design and cybersecurity, to the optimization of grid-enabled and cloud-enabled computational schemes. The advantage is that it offers a vivid cloud platform similar to a real commercial cloud infrastructure. Moreover, it integrates several open source technologies to create an easy-to-use environment, such as Xen, Nimbus, Vine, Hadoop, etc.

DIET.   DIET [DIET, 2011] is a project aiming at implementing distributed scheduling strategies on grids and clouds, initiated by INRIA in 2000.

DIET developed scalable middleware for a multi-agent system, in which clients submit computation requests to a scheduler to find an available server on the grid. In order to facilitate further research in cloud computing, it supplements cloud-specific elements into the scheduler and adds an on-demand resource provision model and an economy-based resource model to test provision heuristics.

SLA@SOI.   SLA@SOI [SLA@SOI, 2011] is a European project aiming at evaluation of service provisioning based on automated SLA management on SOI.

It developed a SLA management framework, which allows the configuration of multi-layer service and automation in an arbitrary service-oriented infrastructure. Besides the scientific values, it implemented a management suite for automated e-contracting and post-sales.

BEinGRID. BEinGRID [BEinGRID, 2011] is a research project providing the infrastructure to support pilot implementations of Grid technologies in actual business scenarios.

In BEinGRID, twenty five business experiments were carried out, each of which focused on a real business problem and the corresponding solution. To extract best practice from the experimental implementations, technical and business consultants worked on analysis of generic components and development of a business plan. Various technologies were evaluated, including cost reduction, enhanced processing power, employing a new business model, running applications such as Software-as-a-Service, etc. Although the BEinGRID project has been completed, it has produced for cloud computing much information regarding requirement knowledge, business drivers, technological solutions and hints for migration potential.

1.5 Cloud challenges

Even though some of the essential characteristics of cloud computing have been realized through commercial and academic efforts, not all capabilities are fulfilled to the necessary extent. Several challenges can be identified as follows

1.5.1 MapReduce programming model

Web servers, Web portals, identity management servers, load balancers and application servers all bring their specific functions to the party for a cloud application. In order to coordinate and use them harmoniously, middleware continues to play a key role in cloud computing. Generally speaking, cloud middleware is the software used to integrate services, applications, and content available on the same or different layers, by which services and other software components can be reused through Internet.

Virtualization is one of the key technologies that can merge different infrastructures, so the management of virtual machines needs to be further developed. Since there is a lot of mature middleware used in grid computing, how can it be combined with cloud middleware? Moreover, a natural evolution from grid to cloud is important, because effort and time can be saved by technology reuse.

As the migration to cloud is inevitable, programming and accessing cloud platforms should perform seamlessly and efficiently. In the future, computational platforms will have a huge number of processing nodes, so traditional parallelization models such as batch processing and message passing models are not scalable enough to deal with large scale distributed computing.

1.5.2   Data management

Data storage in the new cloud platform has changed. Replication and distribution are two of the main characteristics of data stored in a cloud. Data is automatically replicated without the interference of users. Data availability and durability are achieved through replication. Large cloud providers may have datacenters spread across the world.

Data stored in a cloud—replicated and distributed—is considered to be unsuitable for transactional data management applications [Abadi, 2009]. In traditional data management systems, a transaction should support ACID, which means all computations contained in a transaction should behave with Atomicity, Consistency, Isolation and Durability. Such a guarantee is important for write-intensive applications. However, the ACID guarantee is difficult to achieve on replicated and distributed data storage. Among full-fulfilled database products, shared-nothing architecture is not commonly used for transactional data management. Realizing a transaction on a shared-nothing architecture involves complex distributed locking, which is nontrivial work. The advantage of scalability with shared-nothing architecture is not an urgent need for transactional data management.

Analytical data management applications are commonly used in business planning problem solving, and decision support. Data involved in analytical data management is often historical data. Historical data is usually large in size, and is read-mostly (or read-only), and occasionally batch updated. Analytical data management applications can benefit from cloud data storage. Analytical data management applications are argued to be well-suited to run in a Cloud environment [Abadi, 2009], since analytical data management matches well with shared-nothing architecture, and ACID guarantees are not needed for it.

1.5.3   Resource scheduling

Resource management [Magoulès et al., 2008] is the first issue raised over cloud computing platforms. From the provider's point of view, large scale virtual machines need to be allocated to thousands of distributed users, dynamically, fairly, and most importantly, profitably. It's very challenging considering that resource provisioning mechanisms in existing systems such as grids mainly focus on application performance. From the consumer's aspect, users are economy-driven entities when they make the decision to use cloud services [Buyya et al., 2009c]. For adequate resources, one user will compare the price among different providers. For a scarce resource, users themselves becomes competitors who will impact the future price directly, or indirectly. Therefore, the future resource provisioning will become a multi-objective and multi-criteria problem.

For practical reasons, resource provisioning needs reliable and efficient support for negotiation, monitoring, metering, and feedback. The Service Level Agreement (SLA) is a common tool used to define contracts and measure fulfillments in business scenarios. It describes a set of non-functional requirements of the service the customer is buying, and contains penalties when the requirements are not met. Therefore, formal guidelines for contract description have to be standardized.

Besides the technical strengths of cloud computing, users decide to adopt clouds for the economic reasons, so the business model of cloud computing should be more flexible, offering clients scalable price options. For example, Amazon customers can choose purchasing models among on-demand, reserved, spot, and even free tier according to their own preferences. With more and more cloud solutions emerging, business models must be reformed to maintain customer loyalty or attract new attention. In addition, new economic models that support the trading, negotiation, provisioning and allocation based on consumer preference should be developed.

1.6 Concluding remarks

In this chapter, the concept of cloud computing is first introduced. Although there is vast disagreement over what cloud computing is, we tried to refine some representatives and give an unbiased and general definition. That definition is not just an overall concept, but describes system architecture, deployment model and essential features. Cloud computing is still an evolving paradigm, and it integrates many existing technologies. A brief evolution history can help us clarify the conditions, opportunities and challenges existing in cloud development. These definitions, attributes, and characteristics will evolve and change over time.

Functionally speaking, cloud computing is a service provision model, where software, platform, infrastructure, data, and hardware can be directly delivered as a service to end customers. The service characteristics are presented from technical, qualitatives and economic points of view.

Current efforts are the foundation for future research and development. After analyzing existing commercial products and research projects, several challenges in terms of middleware, programming model, resource management and business model are highlighted. These gaps in cloud computing inspire our interest in our future research. In the following chapters, the problem of resource management will be solved in microscopic and macroscopic fashions. In particular, issues such as resource allocation and job scheduling will be studied.

Chapter 2

Resource scheduling for cloud computing

2.1 Introduction

This chapter outlines the problems arising from resource scheduling in cloud computing. Related theories including former expressions of problems, algorithms, complexity and schematic methods are briefly introduced. Then cloud scheduling hierarchy is presented, and scheduling problems split into user-level and system-level. The former focuses on the issue of resource provision between providers and customers, which is solved by economic models. The latter refers to meta-task execution, a suboptimal solution of which is given by heuristics to speed up the process of finding a good enough answer. Moreover, real-time scheduling attracts our attention. Different from economic and heuristic strategies, priority scheduling algorithms and their implementation are discussed at the end of this chapter.

2.2 Cloud service scheduling hierarchy

We specify scheduling problems in cloud environments. As a key characteristic of resource management, service scheduling makes cloud computing different from other computing paradigms. The centralized scheduler in a cluster system aims at enhancing the overall system performance, while the distributed scheduler in a grid system aims at enhancing the performance of specific end-users. Compared with these, scheduling in cloud computing is much more complicated. On the one hand, a centralized scheduler is necessary, because every cloud provider that promises to provide services to users without reference to the hosted infrastructure has an individual datacenter. On the other hand, a distributed scheduler is also indispensable, because commercial property determines that cloud computing should deal with the QoS requirements of customers distributed worldwide. An important issue of this chapter is to decompose scheduling problems related to cloud computing. Since cloud service is actually a virtual product on a supply chain, the service scheduling can be classified into two basic catagories: user-level and system-level. The hierar-

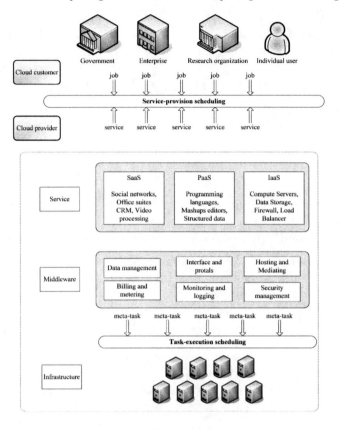

FIGURE 2.1: Scheduling hierarchy.

chy is shown in Figure 2.1. The user-level scheduling deals with the problems raised by service provision between providers and customers. It mainly refers to economic concerns such as equilibrium of supply and demand, competition among consumers and cost minimization under elastic consumer preference. The system-level scheduling handles resource management within a datacenter. From the point of view of customers, a datacenter is an integration system, which provides uniform services. Actually, the datacenter consists of many physical machines, homogeneous or heterogeneous. After receiving numerous tasks from different users, assigning tasks to physical machines significantly impacts the performance of the datacenter. Besides improving the system utilization, some specific requirements should be considered, such as the real-time satisfaction, resource sharing, fault tolerance, etc.

2.3 Economic models for resource-allocation scheduling

In the past three years, the explosion of supply-side cloud service provision has accelerated, with cloud solutions becoming mainstream productions of IT industry. At the same time, these cloud services gradually mature to become more appropriate and attractive to all types of enterprises. The growth of both sides of supply and demand makes the scheduling problems more complex, sophisticated, and even vital in a cloud environment. A bad scheduling scheme not only undermines CPU utilization, turnaround time and cumulative throughput, but may also result in terrible consequences, for example providers losing money and even going out of business.

Economic models are more suitable for cloud-based scheduling than traditional multiprocessor models, especially for regulating the supply and demand of cloud resources. In economics, market-based and auction-based schedulers handle two main interests. Market-based schedulers are applied when a large number of naive users can not directly control service price in commodity trade. Mainstream cloud providers apply market-based pricing schemes in reality. The concrete schemes vary from provider to provider. As the most successful IaaS provider, Amazon EC2 supports commodity and posted pricing models for the convenience of users. Another alternative is the auction-based scheduler, which is adapted to situations where a small number of strategic users seeking to attain a specific service compete with each other. In auctions, users are able to commit the auction price. Amazon spot instance is an example of auction-based model. Instance price adjusts from time to time, depending on the supply and demand. As a result, users should estimate the future price and make their proposal in an auction before placing a spot instance request.

2.3.1 Market strategies

In cloud service provision, both service providers and users express their requirements through SLAs contracts. Providers need mechanisms that support price specification and increase system utilization, while consumers need schemes that guarantee their objectives are reached. A market-based scheduler aims at regulating the supply and demand for resources. To be specific, the market strategies emphasize the schemes for establishing a service price depending on their customers' requirements. In previous literature, a broker behaving on behalf of one end-user interacts with service providers to determine his own price that keeps supply and demand in equilibrium [Buyya et al., 2002].

2.3.1.1 Strategy types

Commodity model. As a common model in our daily life, service providers specify their service price and charge users according to the amount of resource they consume. Any user is free to choose his own provider, but has no right to change the service price directly. The amount of their purchase can cause the price to derive from supply and demand.

The process of scheduling is executed by brokers. On behalf of the users, each broker identifies several providers to inquire about the prices, and then selects one provider which can meet its objective. The consumption of service is recorded and payment is made as agreed.

Posted price model. The posted price strategy makes some special offers to increase the market share or to motivate customers to use the service during the off-peak period. The posted price, as a kind of advertisement, has time or usage limitations that are not suitable for all users. Therefore, the scheduling process should be modified in this strategy.

Service providers give the regular price, the cheap offers and the associated conditions of usage. Brokers observe the posted price, and compare whether it can meet the requirement of users. If not, brokers apply commodity strategy as usual. Otherwise, brokers only inquire of the provider for availability of posted services, supplementing extra regular service when associated conditions are not satisfied.

Bargaining model. In bargaining strategy, the price is not given by the provider unilaterally, but by both sides of the transaction through bargaining. A prerequisite for bargaining is that the objective functions for providers and brokers must have an intersection, so that they can negotiate with each other as long as their objectives are both met. In this scenario, a broker does not compare all the prices for the same service, but connects with one of the providers directly. The price offered by the provider might be higher than customer expectation, so the broker starts with a very low price, which has the upside potential. The bargaining ends when a mutually agreeable price is reached or when one side is not willing to negotiate any further. In the latter case, the broker will connect with other providers and then start bargaining again.

Bargaining strategy has an obvious shortcoming, that is, the overhead on communication is very high. The time delay might lead to lower utilization of resources for the provider or shorten deadline of service for the customers. In reality, the number of negotiations can not be infinite, and the bargaining time is always limited.

2.3.1.2 Principles for strategy design

Several market principles should be considered in the process of determining the service price [Sun et al., 2009].

Equilibrium price. Equilibrium price refers to a price under which the amount of services bought by buyers is equal to the amount of services produced by sellers. This price tends to be stable unless demand or supply change.

Pareto efficiency. Pareto efficiency describes a situation where no agent can get a better allocation than the initial one without reducing other individual allocations. In other words, resources can not be reallocated in a way that makes everyone better off.

Individual rationality. Individual rationality can make prices fluctuate around the equilibrium price, which is determined by the process of supply and demand. A higher price provides incentive to produce more resource, so the amount of scarce resource can gradually reach saturation then surplus, and vice-versa. Individual rationality can adjust prices to reach equilibrium instantaneously.

Stability. Stability examines whether a scheduling mechanism can be manipulated. Individual agents may not reveal private information truthfully. A stable mechanism allows agents to obtain the best allocation if they submit their truthful information.

Communication efficiency. Communication efficiency evaluates the communication overhead to capture a desirable global solution. Message passing adds communication overhead to transactions, so additional time is spent on allocation, rather than on computation. A good scheduling mechanism finds a near-optimum solution efficiently.

2.3.2 Auction strategies

Unlike in market-based models, an auction-based scheduler is a rule maker, rather than a price maker. The rules include how the users bid for services, how the sale price is determined, who the winning bidder is, how the resource is allocated, whether there are limits on time or proposal price, etc.

In auction-based schedulers, price is decided according to the given rules, which benefits consumers by expressing their real requirement strategically, rather than waiting for price adjustment in a passive manner. Auction-based schedulers are distinguished from each other by several characteristics.

2.3.2.1 Strategy types

Number of participants. According to different numbers of bidders, auctions are classified into the demand auction, supply auction and double auction. The English auction is an example of demand auction, in which n buyers bid for one service. This type of auction is the most common form of auction in use today. The Dutch auction focuses on the demand of suppliers, where m sellers offer the same service for one buyer.

Double auction is needed on the condition that the number of buyers and sellers is more than one. In a double auction, sellers and buyers both offer bids. The amount of trade is decided by the quantity at which the marginal buy bid is higher than the marginal sell bid. With the growing number of participants, the double auction converges to the market equilibrium.

Information transparency.   Participants in an auction may or may not know the actions of other participants. Both English and Dutch auctions are open auctions, that is, the participants repeatedly bid for the service with the complete information about previous bids of other bidders. Apart from these, there is another type of auction, in which participants post sealed bids and the bidder with the highest bid wins. In a closed auction, bidders can only submit one bid each and no one knows the other bids. Consequently, blind bidders cannot adjust their bids accordingly.

Closed auction is commonly used for modeling resource provision in multi-agent systems, considering the simplicity and effectiveness of the sealed bids.

Combinatorial auction.   A combinatorial auction is a type of smart market in which participants can place bids on combinations of items, rather than just individual items. Combinatorial auction is appropriate for computational resource auction, where a common procedure accepts bids for a package of items such as CPU cycles, memory, storage, and bandwidth.

Combinatorial auctions are processed by bidders repeatedly modifying their proposals until no one increases their bid any more. In each round, the auctioneer publishes a tentative outcome to help bidders decide whether to increase their bids or not. The tentative outcome is the one that can bring auctioneer the best revenue given the bids. However, finding an allocation of items to maximize the auctioneer's revenue is NP-complete. A challenge of combinatorial auctions comes from how to efficiently determine the allocation once the bids have been submitted to the auctioneer.

Proportion shared auction.   In proportion shared auctions, no winner exists, but all bidders share the whole resource with a percentage based on their bids. This type of auction guarantees a maximized utility and ensures fairness among users in resource allocation, which suits limited resource such as time slot, power, and spectrum bandwidth [Kwok et al., 2005]. Shares represent relative resource rights that depend on the total number of shares contending for a resource. Client allocations degrade gracefully in overload situations, and clients proportionally benefit from extra resources when some allocations are underutilized.

2.3.2.2   Principles for strategy design

Game theoretical equilibrium.   The auction models applied in cloud service and other computational resource provisioning are listed above, but are not limited to these primary types. Generally, an auction-based scheduler emphasizes the equilibrium among users rather than the supply-demand balance between provider and user. The effectiveness of an auction can be analyzed with the help of game theory.

Game theory studies multi-person decision making problems. Any player involved in a game makes the best decision, taking into account the decisions of others. A game theoretical equilibrium is a solution in which no player gains by only changing his own strategy unilaterally. However, this equilibrium does not necessarily mean the best cumulative payoff for all players.

Incentive compatibility. In any auction, participants might hide their true preferences. Incentive compatible auction is one in which participants have incentive to reveal their real private information. A bidder can maximize his payoff only if the information is submitted truthfully.

One method to realize incentive compatibility is designing a reasonable price paid by an auction winner. A good example of incentive compatible auction is the Vickery auction. In this sealed price auction, the highest bidder wins, but pays the second highest bid rather than his own. Under this charging rule, bidding lower or higher than one's true valuation will never increase the best possible outcome.

2.3.3 Economic schedulers

Economic schedulers have been applied to solve resource management in various computing paradigms, such as cluster, distributed databases, grids, parallel systems, Peer-to-Peer, and cloud computing [Buyya et al., 2009c]. Existing middleware applying economic schedulers, not limited to cloud platforms, is introduced. By doing this, we can examine the applicability and suitability of these economic schedulers for supporting cloud resource allocation in practice. This, in turn, helps us identify possible strengths of the middleware that may be leveraged for cloud environment.

Cluster-on-demand. Cluster-on-demand [Cluster-On-Demand, 2011] is a service-oriented architecture for networked utility computing. It creates independent virtual clusters for different groups. These virtual clusters are assigned and managed by a cluster broker, supporting a tendering and contract-net economic model. The user submits its requirements to all cluster brokers. Every broker proposes a specific contract with the estimated execution time and cost. If more than one broker proposes contracts, users then select one as the resource provider. Earning is afforded by users to cluster broker as the cost for adhering to the conditions of the contract.

Mosix. Mosix [Mosix, 2011] is a distributed operating system for high performance cluster computing that employs an opportunity cost approach to minimize the overall execution cost of the cluster. It applies a commodity model to compute a single marginal cost based on the processor and memory usages of the process. The cluster node with the minimal value of marginal cost is then assigned the process.

Stanford Peers. Stanford Peers [Stanford University, 2011] is a peer-to-peer data trading framework, in which both auction and bartering models are applied. A local site wishing to replicate its collection holds an auction to solicit bids from remote sites by first announcing its request for storage space. Each interested remote site then returns a bid, and the site with the lowest bid for maximum benefit is selected by the local site. Besides that, a bartering system supports a cooperative trading environment for producer and consumer participants, so that sites exchange free storage spaces to benefit both themselves and others. Each site minimizes the cost of trading, which is the amount of disk storage space that it has to provide to the remote site for the requested data exchange.

D'Agents.   D'Agents [D'Agents, 2011] is a mobile-agent system for distributed computing. It implements proportion shared auctions where agents compete for shared resources. If there is more than one bidder, resources are allocated proportionally. Costs are defined as rates, such as credits per minute, to reflect the maximum amount that a user wants to pay for the resource.

Nimrod-G.   Nimrod-G [Abramson et al., 2002] is a tool for automated modeling and execution of parameter sweep applications on grids. Through a broker, the grid users obtain service prices from different resources. Deadline and budget are the main constraints specified by the user for running his application. The allocation mechanisms are based on market-based models. Prices of resources thus vary between different executing applications depending on their QoS constraints. A competitive trading environment exists, because users have to compete with one another in order to maximize their own personal benefits.

Faucets.   Faucets [Kale et al., 2004] is a resource scheduler of computational grid, and its objective is supporting efficient resource allocation for parallel jobs executed on a changing number of allocated processors during runtime on demand. A tendering model is used in Faucets. A QoS contract is agreed to before job execution, including payoff at soft deadline, a decreased payoff at hard deadline and a penalty after hard deadline. Faucets aims to maximize the profit of resource provider and resource utilization.

MarketNet.   MarketNet [Dailianas et al., 2000] is a market-based protection technology for distributed information systems. A posted price model is incorporated. Currency accounts for information usage. The MarketNet system advertises resource request by offering prices on a bulletin board. Through observing currency flow, potential intrusion attacks into the information systems are controlled, and the damage is kept to the minimum.

Cloudbus.   Cloudbus [Buyya et al., 2009a] is a toolkit providing market-based resource management strategies to mediate access to distributed physical and virtual resources. A third party cloud broker is built on an architecture that provides a general framework for any other cloud platforms. A number of economic models with commodity, tendering, and auction strategies are available for customer-driven service management and computational risk management. The broker supports various application models such as parameter sweep, workflow, parallel, and bag of tasks. It has plug-in support for integration with other middleware technologies such a Globus, Aneka, Unicore, etc.

OpenPEX. OpenPEX [Venugopal et al., 2009] is a resource provisioning system with an advanced reservation approach for allocating virtual resources. A user can reserve any number of instances of virtual machines that have to be started at a specific time and have to last for a specific duration. A bilateral negotiation protocol is incorporated in OpenPEX, allowing users and providers to exchange their offers and counteroffers, so more sophisticated bartering or double auction models are helpful to increase the revenue of cloud users.

EERM. EERM [Elmroth and Tordsson, 2005] is a resource broker that enables bidirectional communication between business and resource layers to promote good decision-making in resource management. EERM contains sub-components for performing pricing, accounting, billing, job-scheduling, monitoring and dispatching. It uses all kinds of market-based mechanisms for allocating network resources. To increase the revenue, overbooking strategy is implemented to mitigate the effects of cancellations and no-shows.

A summary of economic schedulers is concluded in Table 2.3.3.

Table 2.1: Economic Schedulers.

Scheduler	Economic model	Computing paradigm
Cluster-on-demand	tendering	cluster
Mosix	commodity	cluster
Stanford Peers	auction/bartering	peer to peer
D'Agents	proportion shared auction	mobile-agent
Faucets	tendering	grid
Nimrod-G	commodity/auctions	grid
MarketNet	posted price	distributed information
Cloudbus	commodity/tendering/auctions	cloud
OpenPEX	bartering/double auction	cloud
EERM	commodity/posted price/... ... /bartering/tendering	cloud

2.4 Heuristic models for task-execution scheduling

In cloud computing, a typical datacenter consists of commodity machines connected by high-speed links. This environment is well suited to the computation of large, diverse groups of tasks. Tasks belonging to different users are no longer distinguished one from another. The scheduling problem in such a context is matching multi tasks to multi machines. As mentioned in the previous section, the optimal matching is an optimization problem, generally with NP-complete complexity. Heuristics are often applied as a sub-optimal algorithm to obtain relatively good solutions.

This section intensively researches two types of strategies, static and dynamic heuristics. A static heuristic is suitable for the situation where the complete set of tasks is known prior to execution, while a dynamic heuristic performs the scheduling when a task arrives. Before further explanation, several preliminary terms should be defined.

- t_i: task i

- m_j: machine j

- c_i: the time when task t_i arrives

- a_j: the time when machine m_j is available

- e_{ij}: the execution time for t_i is executed on m_j

- c_{ij}: the time when the execution of t_i is finished on m_j, $c_{ij} = a_j + e_{ij}$

- *makespan*: the maximum value of c_{ij}, which means the whole execution time. The aim of heuristics is to minimize makespan, that is to say, scheduling should finish execution of a metatask as soon as possible.

2.4.1   Static strategies

Static strategies are performed under two assumptions. The first is that tasks arrive simultaneously $c_i = 0$. The second is that machine available time a_j is updated after each task is scheduled.

Opportunistic Load Balancing (OLB).   OLB schedules every task, in arbitrary order, to the next available machine. Its implementation is quite easy, because it does not need extra calculation. The goal of OLB is simply to keep all the machines as busy as possible.

Minimum Execution Time (MET).   MET schedules every task, in arbitrary order, to the machine which has the minimum execution time for this task. MET is also very simple, giving the best machine to each task, but it ignores the availability of machines. MET jeopardizes the load balance across machines.

Minimum Completion Time (MCT).   MCT schedules every task, in arbitrary order, to the machine which has the minimum completion time for this task. However, in this heuristic, not all tasks can be given the minimum execution time.

Min-min.   Min-min begins with the set T of all unscheduled tasks. Then, the matrix for minimum completion time for each task in set T is calculated. The task with overall minimum completion time is scheduled to its corresponding machine. Next, the scheduled task is removed from T. The process repeats until all tasks are scheduled.

Min-max.   Min-max is similar to Min-min heuristic. Min-max also begins with the set T of all unscheduled tasks, and then calculates the matrix for minimum completion time for each task in set T. Different from min-min, the task with overall maximum completion time is selected and scheduled to its corresponding machine. Next, the scheduled task is removed from T. The process repeats until all tasks are scheduled.

Genetic Algorithm (GA).   GA is a heuristic to search for a near-optimal solution in large solution spaces [Braun et al., 2001]. The first step is randomly initializing a population of chromosomes (possible scheduling) for a given task. Each chromosome has a fitness value (makespan) that results from the scheduling of tasks to machines within that chromosome. After the generation of the initial population, all chromosomes in the population are evaluated based on their fitness value, with a smaller makespan being a better mapping. Selection scheme probabilistically duplicates some chromosomes and deletes others, where better mappings have a higher probability of being duplicated in the next generation. The population size is constant in all generations. Next, the crossover operation selects a random pair of chromosomes and chooses a random point in the first chromosome. The crossover exchanges machine assignments between corresponding tasks. A mutation operation is performed after the crossover. Mutation randomly selects a chromosome, then randomly selects a task within the chromosome, and randomly reassigns it to a new machine. After evaluating the new population, another iteration of GA starts, including selection, crossover, mutation, and evaluation. Only when stopping criteria are met, will the iteration stop.

Simulated Annealing (SA).   SA uses a procedure that probabilistically allows poorer solutions to be accepted to obtain a better search of the solution space. This probability is based on a system temperature that decreases for each iteration, which implies that a poorer solution is not easily accepted. The initial system temperature is the makespan of the initial scheduling, which is mutated in the same manner as the GA. The new makespan is evaluated at the end of each iteration. A worse makespan might be accepted based on a probability, so the SA finds poorer solutions than min-min and GA.

Tabu.   A Tabu search keeps track of the regions of the solution space which have already been searched so as not to repeat a search near these areas. A scheduling solution uses the same representation as a chromosome in the GA approach. To manipulate the current solution and to move through the solution space, a short hop is performed. The intuitive purpose of a short hop is to find the nearest local minimum solution within the solution space. When the short hop procedure ends, the final scheduling from the local solution space search is added to the Tabu list. Next, a new random scheduling is generated, to perform a long hop to enter a new unsearched region of the solution space. After each successful long hop, the short hop procedure is repeated. After the stopping criterion is satisfied, the best scheduling from the Tabu list is the final answer.

A^*. A^* is a tree-based search heuristic beginning at a root node that is a null solution. As the tree grows, nodes represent partial scheduling (a subset of tasks is assigned to machines), and leaves represent final scheduling (all tasks are assigned to machines). The partial solution of a child node has one more task scheduled than the parent node. Each parent node can be replaced by its children. To keep execution time of the heuristic tractable, there is a pruning process to limit the maximum number of active nodes in the tree at any one time. If the tree is not pruned, this method is equivalent to an exhaustive search. This process continues until a leaf (complete scheduling) is reached.

The listed heuristics above are fit for different scheduling scenarios. The variation of scenarios is caused by the task heterogeneity, machine heterogeneity and machine inconsistence. The machines are consistent if machine m_i executes any task faster than machine m_j, it executes all tasks faster than m_j. These heuristics are evaluated by simulation in an article [Braun et al., 2001]. For consistent machines, GA performs the best, while MET performs the worst. For inconsistent machines, GA and A^* give the best solution, and OLB gives the worst. Generally, GA, A^* and min-min can be used as a promising heuristic with a short average makespan.

2.4.2   Dynamic strategies

Dynamic heuristics are necessary when a task set or a machine set is not fixed. For example, not all tasks arrive simultaneously, or some machines go offline at intervals. The dynamic heuristics can be used in two fashions, on-line mode and batch mode. In the former mode, a task is scheduled to a machine as soon as it arrives. In the latter mode, tasks are first collected into a set that is examined for scheduling at prescheduled times.

2.4.2.1   On-line mode

In on-line heuristics, each task is scheduled only once; the scheduling result can not be changed. On-line heuristics are suitable for the cases in which arrival rate is low [Shoukat et al., 1999].

OLB.   The OLB dynamic heuristic assigns a task to the machine that becomes ready next regardless of the execution time of the task on that machine.

MET. A MET dynamic heuristic assigns each task to the machine that performs that task's computation in the least amount of execution time, regardless of machine available time.

MCT. The MCT dynamic heuristic assigns each task to the machine, which results in the task's earliest completion time. MCT heuristic is used as a benchmark for the on-line mode [Shoukat et al., 1999].

Switching Algorithm (SA). The SA uses the MCT and MET heuristics in a cyclic fashion depending on the load distribution across the machines. MET can choose the best machine for tasks but might assign too many tasks to the same machines, while MCT can balance the load, but might not assign tasks to machines that have their minimum executing time. If the tasks are arriving in a random mix, it is possible to use the MET at the expense of load balance up to a given threshold, and then use the MCT to smooth the load across the machines.

K-Percent Best (KPB). The KPB heuristic considers only a subset of machines while scheduling a task. The subset is formed by picking the k best machines based on the execution times for the task. A good value of k schedules a task to a machine only within a subset formed from computationally superior machines. The purpose is to avoid putting the current task onto a machine which might be more suitable for some yet-to-arrive tasks, so it leads to a shorter makespan as compared to the MCT.

For all the on-line mode heuristics, KPB outperforms others in most scenarios [Shoukat et al., 1999]. The results of MCT are good, only slightly worse than KPB, owing to the lack of prediction for task heterogeneity.

2.4.2.2 Batch mode

In batch mode, tasks are scheduled only at some predefined moments. This enables batch heuristics to determine about the actual execution times of a larger number of tasks.

Min-min. First Min-min updates the set of arrival tasks and the set of available machines, calculating the corresponding expected completion time for all ready tasks. Next, the task with the minimum earliest completion time is scheduled and then removed from the task set. Machine available time is updated, and the procedure continues until all tasks are scheduled.

Max-min. The Max-min heuristic differs from the Min-min heuristic where the task with the maximum earliest completion time is determined and then assigned to the corresponding machine. The Max-min performs better than the Min-min heuristic, if the number of shorter tasks is larger than that of longer tasks.

Sufferage. The Sufferage heuristic assigns a machine to a task that would suffer most if that particular machine were not assigned to it. In every scheduling event, a sufferage value is calculated, which is the difference between the first and the second earliest completion time. For task t_k, if the best machine m_j with the earliest completion time is available, t_k is assigned to m_j. Otherwise, the heuristic compares the sufferage value of t_k and t_i, the task already assigned to m_j. If the Sufferage value of t_k is bigger, t_i is unassigned and added back to the task set. Each task in the set is considered only once.

Generally, Sufferage gives the smallest makespan among batch mode heuristics [Shoukat et al., 1999]. The batch mode performs better than the on-line mode with high task arrival rate.

2.4.3   Heuristic schedulers

One advantage of cloud computing is that tasks which might be difficult, time consuming, or expensive for an individual user can be efficiently accomplished in a datacenter. Datacenters in clouds support functional separation between the processing power and data storage, both of which locate in a large number of remote devices. Hence, scheduling becomes more complicated and challenging than ever before. Since a scheduler is only a basic component of the whole infrastructure, no general scheduler can fit for all cloud architectures. In this section, we mainly discuss schedulers used for data-intensive distributed applications.

2.4.3.1   Hadoop

MapReduce is a popular computation framework for processing large-scaled data in mainstream public and private clouds, and it is considered an indispensable cornerstone for cloud implementation. Hadoop is the most widespread MapReduce implementation for educational or production uses. It enables applications to work with thousands of nodes and petabytes of data.

A multi-node Hadoop cluster contains two layers. The bottom is the Hadoop Distributed File System (HDFS), which provides data location awareness for effective scheduling of work. Above the file systems is the MapReduce engine, which includes one job tracker and several task trackers. Every tracker inhabits an individual node. Clients submit MapReduce jobs to the job tracker, then the job tracker pushes work out to available Task Tracker nodes in the cluster [Borthakur, 2007].

Hadoop is designed for large batch jobs. The default scheduler uses an FIFO heuristic to schedule jobs from a work queue. Alternative job schedulers are the fair scheduler, the capacity scheduler, and the delay scheduler.

FIFO scheduler. The FIFO scheduler [Borthakur, 2007] applies a first-in first-out heuristic. When a new job is submitted, the scheduler puts it in the queue according to its arrival time. The earliest job on the waiting list is always executed first. The advantages are that the implementation is quite easy and that the overhead is minimal. However, throughput of FIFO scheduler is low, since tasks with long execution time can seize the machines.

Fair scheduler. A Fair scheduler [Zaharia et al., 2008] assigns an equal share of resources to all jobs. When new jobs are submitted, task slots that free up are shared, so that each job gets roughly the same amount of CPU time. The Fair scheduler supports job priorities as weights to determine the fraction of total compute time that each job should be assigned. It also allows a cluster to be shared among a number of users. Each user is given a separate pool by default, so that everyone is allocated the same share of the cluster no matter how many jobs are submitted. Within each pool, fair sharing is used to share capacity between the running jobs. In addition, a guaranteed minimum share is allowed. When a pool contains jobs, it gets at least its minimum share, but when the pool does not need its full guaranteed share, the excess is split among other running jobs.

Capacity scheduler. A Capacity scheduler [Zaharia et al., 2009] allocates cluster capacity to multiple queues, each of which contains a fraction of capacity. Each job is submitted to a queue; all jobs submitted to the same queue will have access to the capacity allocated to the queue. Queues enforce limits on the percentage of resources allocated to a user at any given time, so no user monopolizes the resource. Queues optionally support job priorities. Within a queue, jobs with high priority will have access to resources preferentially. However, once a job is running, it will not be preempted for a higher priority job.

Delay scheduler. A Delay scheduler [Zaharia et al., 2010] addresses conflict between scheduling fairness and data locality. It temporarily relaxes fairness to improve locality by asking jobs to wait for a scheduling opportunity on a node with local data. When the job that should be scheduled next according to fairness cannot launch a local task, it waits for a short length of time, letting other jobs launch tasks instead. However, if a job has been skipped long enough, it is allowed to launch non-local tasks to avoid starvation. A Delay scheduler is effective if most of the tasks are short compared to jobs, and if there are many slots per node.

2.4.3.2 Dryad

The Dryad [Dryad, 2011] is a distributed execution engine for general data parallel applications, and it seems to be Microsoft's programming framework, providing similar functionality to Hadoop. Dryad applies directed acyclic graph (DAG) to model applications.

Quincy. The Quincy [Isard et al., 2009] scheduler tackles the conflict between locality and scheduling in Dryad framework. It represents the scheduling problem as an optimization problem. Min-cost flow makes a scheduling decision, matching tasks and nodes. The basic idea is to kill some of the running tasks and then to launch new tasks to place the cluster in the configuration returned by the flow solver.

2.4.3.3   Others

To sum up the heuristic schedulers for cloud computing, scheduling in clouds is all about resource allocation, rather than job delegation in HPC or grid computing. However, the traditional meta-schedulers can be evolved to adapt cloud architectures and implementations, considering the development of virtualization technologies. Next, we take several representatives for example as follows

Oracle Grid Engine.   The Oracle Grid Engine [Oracle, 2011] is an open source batch-queuing system. It is responsible for scheduling remote execution of large numbers of standalone, parallel or interactive user jobs and managing the allocation of distributed resources. It is now integrated by Hadoop and Amazon EC2, and works as a virtual machine scheduler for Nimbus in a cloud computing environment.

Maui Cluster Scheduler.   The Maui Cluster Scheduler [Maui, 2011] is an open source job scheduler for clusters and supercomputers, which is capable of supporting an array of scheduling policies, dynamic priorities, extensive reservations, and fair share capabilities. It has now developed new features including virtual private clusters, basic trigger support, graphical administration tools, and a Web-based user portal in Moab.

Condor.   Condor [Thain et al., 2005] is an open source high-throughput computing software framework used to manage workload on a dedicated cluster of computers. Condor-G has developed, provisioning virtual machines on EC2 through the VM Universe. It also supports launching Hadoop MapReduce jobs in Condor's parallel universe.

gLite.   gLite [Ragusa et al., 2009] is a middleware stack for grid computing initially used in scientific experiments. It provides a framework for building grid applications, tapping into the power of distributed computing and storage resources across the Internet, which can be compared to corresponding cloud services such as Amazon EC2 and S3. Since technologies such as REST, HTTP, hardware virtualization and BitTorrent displaced existing accesses to grid resources, gLite federates both resources from academic organizations as well as commercial providers to remain pervasive and cost effective.

2.5 Real-time scheduling in cloud computing

There are emerging classes of applications that can benefit from increasing the timing guarantee of cloud services. These mission critical applications typically have deadline requirements, and any delay is considered a failure for the whole deployment. For instance, traffic control centers periodically collect data on the state of roads using sensor devices. Database updates recent information before next data reports are submitted. If anyone consults the control center about traffic problems, a real-time decision should be made to help operators choose appropriate control actions. Besides, current service level agreements can not provide cloud users with real-time control over the timing behavior of the applications, so more flexible, transparent and trustworthy service agreements between cloud providers and users are needed in future.

Given the above analysis, the ability to satisfy timing constraints of such real-time applications plays a significant role in the cloud environment. However, the existing cloud schedulers are not perfectly suitable for real-time tasks, because they lack strict requirement of hard deadlines. A real-time scheduler must ensure that processes meet deadlines, regardless of system load or makespan.

Priority is applied to the scheduling of these periodic tasks with deadlines. Every task in priority scheduling is given a priority through some policy, so that scheduler assigns tasks to resources according to priorities. Based on the policy for assigning priority, real-time scheduling is classified into two types: fixed priority strategy and dynamic priority strategy.

2.5.1 Fixed priority strategies

A real-time task τ_i contains a series of instances. Fixed priority scheduling is when all instances of one task have the same priority. The most influential algorithm for priority assignment is the Rate Monotonic (RM) algorithm proposed by Liu [Liu and Layland, 1973]. In the RM algorithm, the priority of one task depends on its release rate. The higher the rate is, the higher the priority. Period T_i is the length of time between two successive instances, and computation time C_i is the time spent on task execution. Since the release rate is inverse to its period, T_i is usually the direct criterion to determine task priority.

A schedulability test is to determine whether temporal constraints of tasks can be met at runtime. Exact tests are ideal but intractable, because the complexity of exact tests is NP-hard for non-trivial computational models [Sha et al., 2004]. Sufficient tests are less complex but more pessimistic. Schedulability analysis is suitable for the systems whose tasks are known *a priori*.

Sufficient tests can be executed by checking whether a sufficient utilization-based condition is met. For example, Liu [Liu and Layland, 1973] proved that a set of n periodic tasks using RM algorithm is schedulable if $\sum \frac{C_i}{T_i} \leq n(2^{1/n} - 1)$. The bound is tight, in the sense that some task sets are unschedulable with the utilization that is arbitrarily higher than $n(2^{1/n} - 1)$. Actually, many task sets with utilization higher than this bound can be scheduled. Lehoczky [Lehoczky et al., 1989] proved that the average schedulable utilization, for large randomly chosen task sets, reaches 0.88, much higher than Liu's result of 0.69. The desire for a more precise and tractable schedulability test pushes researchers to search for high utilization bounds under special assumptions, such as appropriate choice of task periods.

Exact testing permits higher utilization levels to be guaranteed. One approach to solving this problem is determining the worst-case response time of a task R_i. Once the longest time between arrival of a task and its subsequent instantiations is known, the test can be checked by comparing the deadline D_i and the worst-case response time R_i. The complexity of the test comes from the R_i calculation by recursive equations. $R_i = C_i + \sum_{j=1}^{i-1} \left\lceil \frac{R_i}{T_j} \right\rceil C_j$. This equation can be solved iteratively, because only a subset of the task release times in the interval between zero and T_i needs to be examined, observed by Harter, Joseph and Audsley independently [Harter and Paul, 1987, Joseph and Pandya, 1986, Audsley et al., 1993].

One relaxation of Liu's model is that task deadline does not exactly equal its period. Therefore, a RM algorithm is not optimal for priority assignment. Instead, Leung proposed the Deadline Monotonic (DM) algorithm as the optimal policy for such systems, assigning higher priorities to tasks with shorter deadlines than those with longer deadlines [Leung and Whitehead, 1982]. Under this assumption, Lehoczky [Lehoczky, 1990] proposed two sufficient schedulability tests by restricting $D_i = kT_i$, where k is a constant across all tasks. Tindell [Thuel and Lehoczky, 1994] extended the exact test for tasks with arbitrary deadlines.

A further relaxation is permitting tasks to have unequal offsets. Since the worst-case situation occurs when all tasks share a common release time, utilization bound for sufficient test and response time for exact test in Liu's model might be too pessimistic. Analyzing general offsets efficiently still remains a problem. Under the assumption of specified offsets, RM and DM are no longer optimal, but Audsley [Audsley et al., 1995] showed the optimal priority assignment can be achieved by examining a polynomial number of priority ordering over the task set.

Liu's model and its further extensions are suitable for single processor scheduling. In distributed systems, multiple processors can be scheduled in two approaches, partitioned and global. The former is when each task is assigned to one processor, which executes all incantations of the task. The latter is when tasks compete for the use of all processors. Partition and global schemes are incomparable in effectiveness, since the required number of processors is not the same [Sha et al., 2004].

For partitioned policy, the first challenge is to find the optimal partitioning of tasks among processors, which is a NP-complete problem. Therefore, heuristics are used to find good sub-optimal static allocations. The main advantage of heuristic approaches is that they are much faster than optimal algorithms, while they deliver fairly good allocations. Dhall [Dhall and Liu, 1978] proved that RM Next-Fit guarantees schedulability of task sets with utilization bound of $m/(1 + 21/3)$. Oh [Oh and Bakker, 1998] showed that RM First-Fit schedules periodic tasks with total utilization bounded by $m(21/2 - 1)$. Later, Lopez [López et al., 2004] lifted a tight bound of $(m+1)(2^{1/(m+1)} - 1)$ for RM First-Fit scheduling. Andersson [Andersson et al., 2001] showed that system utilization can not be higher than $(m+1)/2$ for any combination of processor partitioning and any priority assignment.

For global policy, the greatest concern is to find an upper bound λ on the individual utilization for RM global scheduling. The small λ presents high system utilization bound. Andersson [Andersson et al., 2001] proved that system utilization bound is $m^2/(3m - 1)$ with $\lambda = m/(3m - 2)$. Baruah [Baruah and Goossens, 2003] showed that for $\lambda = 1/3$ system utilization of at least $m/3$ can be guaranteed. With arbitrary large λ, Barker [Baker, 2005] showed that the system utilization bound is $(m/2)(1 - \lambda) + \lambda$.

2.5.2 Dynamic priority strategies

Dynamic priority assignment is more efficient than the fixed manner, since it can fully utilize the processor for the most pressing tasks. The priorities change with time, varying from one request to another or even during the same request. The most used algorithms are Earliest Deadline First (EDF) and Least Laxity First (LLF) [Uthaisombut, 2008]. EDF assigns priorities to tasks inversely proportional to the absolute deadlines of the active jobs. Liu [Liu and Layland, 1973] proved that n periodic tasks can be scheduled using the EDF algorithm if and only if $\sum \frac{C_i}{T_I} \leq 1$. LLF assigns the processor to the active task with the smallest laxity. LLF has a large number of context switches due to laxity changes at runtime. Even though both EDF and LLF are optimal algorithms, EDF is more popular in real-time research because it has a lower overhead than LLF.

Under EDF, schedulability tests can be done by processor demand analysis. Processor demand in an interval $[t_1, t_2]$ is the amount of processing time $g(t_1, t_2)$ requested by those tasks that must be completed in $[t_1, t_2]$. The tasks can be scheduled if and only if any interval of time the total processor demands $g(t_1, t_2)$ is less than the available time $[t_1, t_2]$. Baruah [Braun et al., 2001] proved that a set of periodic tasks with the same offset can be scheduled if and only if $U < 1$ and $\forall L > 0, \sum_{i=1}^{n} \left\lfloor \frac{L + T_i - D_i}{T_i} \right\rfloor C_i \leq L$. The sufficient test of EDF is of $O(n)$ complexity if deadline equals period. Otherwise, an exact test can be finished in pseudo-polynomial time complexity, when deadline is no longer than period [Sha et al., 2004].

The research on real-time scheduling is not limited to the issues discussed above. For practicable usage, assumptions can be released, so that researches are extended in a number of ways.

- Not all the tasks have periodic release. An aperiodic server is introduced to permit aperiodic tasks to be accommodated in the periodic models.

- Tasks have resource or precedence relationships. Tasks can be linked by a linear precedence constraint, and communicating via shared resources is allowed to realize task interaction.

- Computation time of tasks varies widely. Some reduced-but-acceptable level of service should be provided when workload exceeds normal expectations.

- Soft real-time applications exist. Control mechanisms can optimize the performance of the systems, and analytic methods are developed to predict the system performance.

2.5.3   Real-time schedulers

A scheduler is called dynamic if it makes scheduling decisions at runtime, selecting one out of the current set of ready tasks. A scheduler is called static (pre-runtime) if it makes scheduling decisions at compile time. A static scheduler generates a dispatching table for the runtime dispatcher off-line.

Generally, real-time schedulers are embedded in corresponding kernels with respect to their scheduling approaches. The MARS kernel [Hyman et al., 1991] targets hard real-time systems for peak load conditions. A fixed scheduling approach is adopted. Schedule is completely calculated offline and is given to the nodes as part of system initialization. All inter-process communications and resource requests are included in the schedule. Nodes may change schedules simultaneously to another pre-calculated schedule.

An arts kernel [Tokuda and Mercer, 1989] aims at providing a predictable, analyzable, and reliable distributed computing system. It uses the RM/EDF/LLF algorithms to analyze and guarantee hard real-time processes offline. Non-periodic hard real-time processes are scheduled using execution time reserved by a deferrable server. All other processes are scheduled dynamically using a value-function scheme.

With the augmentation of real-time services, real-time kernels are widely required in cloud computing. However, many kernels are not very capable of satisfying real-time systems requirements, particularly in the multicore context. One solution is applying loadable real-time schedulers as plug-ins into operation systems regardless of kernel configurations. As a result, variant scheduling algorithms are easily installed. A good example is RESCH for the Linux kernel, which implements four scheduler plugins with partitioned, semi-partitioned, and global scheduling algorithms [Kato et al., 2009].

When schedulers step into the cloud environment, virtualization is an especially powerful tool. Virtual machines can schedule real-time applications [Buyya et al., 2009a], because they allow for a platform-independent software development and provide isolation among applications. For example, Xen provides the simplest EDF scheduler to enforce temporal isolation among the different VMs. OpenVMS, a multi-user multiprocessing virtual memory-based operating system, is also designed for real-time applications.

2.6 Concluding remarks

In this chapter, we first reviewed the scheduling problems in a general fashion. Then we described the cloud service scheduling hierarchy. The upper layer deals with scheduling problems raised by economic concerns, such as equilibrium between service providers and consumers, the competition among consumers who need the same service, etc. Market-based and auction models are effective tools, both of which are explained with details and design principles. After that, several types of middleware leveraging these economic models for cloud environment are presented. The lower layer refers to metadata scheduling within the datacenter. Tasks belonging to different users are no longer distinguished one from another. The scheduling problem in such a context is matching multi tasks to multi machines, which can be solved by heuristics. Heuristics are classified into two types. A static heuristic is suitable for the situation where the complete set of tasks is known prior to execution, while a dynamic heuristic performs the scheduling when the task arrives. In cloud-related frameworks, such as Hadoop and Dryad, batch-mode dynamic heuristics are most used, and more practical schedulers are developed for special usage. Other meta-schedulers in HPC or grid computing have evolved to adapt cloud architectures and implementations.

For commercial purposes, cloud services heavily emphasize time guarantee. The ability to satisfy timing constraints of such real-time applications plays a significant role in the cloud environment. We then examined the particular scheduling algorithms for real-time tasks, that is, priority-based strategies. These strategies, already used in traditional real-time kernels, are not really capable of satisfying real-time systems requirements. New technologies, such as loadable real-time plug-ins and virtual machines, are introduced as promising solutions for real-time cloud schedulers.

Chapter 3

Game theoretical allocation in a cloud datacenter

3.1 Introduction

Cloud computing can cut IT costs and at the same time herald in a new era of agility in IT operations. A fundamental element is the concept of a datacenter, in which IT solutions are considered as services and are as easily purchased as other consumption models. This facility is caused by the development of virtualization technology, which hides heterogeneous configuration details from customers. Therefore, resource provision takes on market dealing behaviors, not just match-making scheduling between tasks and machines [Armbrust et al., 2009]. The market mechanism is an effective method to control electronic resources, but the existing market-based models are dedicated either to maximize the revenue of suppliers, or to balance the supply-demand relationship [Buyya et al., 2002]. In this chapter, we shall focus on the contest among cloud customers who need the same resource, and make a reasonable allocation to keep market equilibrium.

Game theory studies multi-person decision making problems. Although there have been researches on allocation strategies using game theory [Galstyan et al., 2003], [Bredin et al., 2003], [Maheswaran and Basar, 2003], [Khan and Ahmad, 2006], [An et al., 2007], [Wei et al., 2009], [Fan et al., 2009], [Teng and Magoulès, 2010], no one strategy perfectly suits the new computing service market. In order to establish an appropriate model for clouds, several important characteristics should be highlighted. First, cloud users, whose goal is to get better service at a better cost, are self-interested but rational. Second, these buyers have more than one behavioral constraint, so they have to make a trade-off of one constraint for another in management practice. Third, the pay-as-you-go feature means transactions are never static, but repeated gambling processes. Each user can adjust its bid price based on the previous behaviors. Fourth, cloud customers are anonymous, in different cities worldwide, so they do not know each other. In other words, there is no common procurement knowledge in the whole system. Fifth, cloud users having different tasks always arrive in datacenters without a prior arrangement, in which, the accurate forecast becomes extremely challenging in such a complex scenario, so a good allocation model integrating compromise, competition and prediction should be further generalized and well evaluated. Given the above challenges, we thereby use game theoretical auctions to solve the resource allocation problem in clouds, and propose practicable algorithms for user bidding and auctioneer pricing. With Bayesian learning prediction, resource allocation can reach Nash equilibrium among non-cooperative users even though common knowledge is insufficient and dynamically updated.

The remainder of this chapter is organized as follows. Section two first gives a short tutorial on game theory, covering the different classes of games and their applications, payoff choice and utility function, as well as strategic choice and Nash equilibrium. Next, a non-cooperative game for resource allocation is built. The scheduling model includes bid-shared auction, user strategy (bid function), price forecasting and equilibrium analysis. Based on equilibrium allocation, simulation algorithms running on the Cloudsim platform are proposed. After that, the Nash equilibrium and forecasting accuracy are evaluated. We conclude this chapter by summarizing related work on game theoretical resource allocation and by suggesting some future research avenues.

3.2   Game theory

Game theory models strategic situations, in which an individual's payoff depends on the choices of others. It provides a theoretical basis for the fields of economics, business, politics, logic, and computer science. It is an effective approach to achieve equilibrium in multi-agent systems, computational auctions, peer-to-peer systems, and security and information markets. With the development of the cloud service market, game theory is useful to address the resource allocation problems in cloud systems where agents are autonomous and self-interested.

3.2.1   Normal formulation

Game is an interactive environment where the benefit of an individual choice depends on the behaviors of other competitors. A normal game consists of all conceivable strategies of every player and their corresponding payoffs. There are several important terms used to characterize a normal form of game [Gibbons, 1992].

Player.   A player is the game participant. There is a finite set of players $P = \{1, 2, \cdots, m\}$.

Strategy.   Strategy is the action taken by one player. Each player k in P has a particular strategy space containing a finite number of strategies, $S_k = \{s_k^1, s_k^2, \cdots, s_k^n\}$. Strategy space is $S = S_1 \times S_2 \times \cdots \times S_m$. The game outcome is a combination of strategies of m players $s = (s_1, s_2, \cdots, s_m), s_i \in S_i$.

Payoff. Payoff is the utility received by a single player at the outcome of one game, which determines the player's preference. For resource allocation, payoff stands for the amount of resource received, for example, $u_i(s)$ represents the payoff of player i when the output of the game is $s, s \in S$. A payoff function $U = \{u_1(S), u_2(S), \cdots, u_m(S)\}$ is specified for each player defined by the player set P.

Therefore, the normal form of a game is a structure such as

$$G = < P, S, U > \tag{3.1}$$

3.2.2 Payoff choice and utility function

In the cloud computing market, service providers and their customers have their own preferences. Providers balance the investments on capital, operation, labor, and device. Customers have different QoS requirements, such as cost, execution time, access speed, throughput, and stability. All these preferences have an impact on agents' choices, thus an integrated indication to guide agents' behaviors is necessary.

Utility is a measure of relative satisfaction in economics. It is often expressed as a function to describe the payoff of agents. More specifically, utility function combines more than one service requirement and analyzes Pareto efficiency under certain assumptions such as service consumption, time spending, and money possession. Therefore, utility is very useful when a cloud agent tries to make a wise decision. High value of utility stands for great preference of service when the inputs are the same.

One key property of utility function is constant elasticity of substitution (CES). It combines two or more types of consumption into an aggregate quantity. The CES function is

$$C = [\sum_{i=1}^{n} a_i^{\frac{1}{s}} c_i^{\frac{s-1}{s}}]^{\frac{s}{s-1}} \tag{3.2}$$

C is aggregate consumption, c_i is individual consumptions, such as energy, labor, time, capital, etc. The coefficient a_i is share parameter, and s is elasticity of substitution. These consumptions are perfect substitutes when s approaches infinity, and are perfect complements when s approaches zero. The preferences for one factor over another always change, so the marginal rate of substitution is not constant. For the sake of simplicity, s equals one in the following analysis. Let $r = (s-1)/s = 0$, we obtain

$$\ln C = \frac{\ln \sum_{i=1}^{n} (a_i^{1-r} c_i^r)}{r} \tag{3.3}$$

Apply l'Hopital's rule,

$$\lim_{r \to 0} \ln C = \frac{\sum_{i=1}^{n} a_i \ln c_i}{\sum_{i=1}^{n} a_i} \tag{3.4}$$

If $\sum_{i=1}^{n} a_i = 1$, the consumption function has constant returns to scale, which means that the consumption increased by the same percentage as the rate of growth of each consumption good. If every a_i is increased by 20%, C increases by 20% accordingly. If $\sum_{i=1}^{n} a_i < 1$, the returns to scale decrease, on the contrary, returns to scale increase. We take two QoS requirements, speed and stability, for example. The CES function is shown in Figure 3.1. The contour plot beneath the surface signifies a

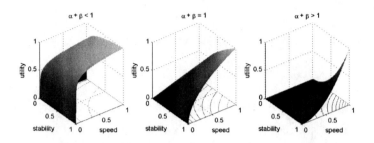

FIGURE 3.1: Constant elasticity of substitution functions.

collection of indifference curves, which can represent observable demand patterns over the right bundles. Every curve shows different bundles of goods, for which a consumer has no preference for one bundle over another. One can equivalently refer to each point on the indifference curve as rendering the same level of utility for the customer.

In particular, CES function is a general expression of the Cobb Douglas function. The Cobb Douglas function has been widely used in consumption, production and other social welfare analysis. It can build a utility function. In a generalized form, where $c_1, c_2, \cdots, c_n$ are the quantities consumed of n goods, the utility function representing the same preferences is written as:

$$\tilde{u}(c) = \prod_{i=1}^{n} c_i^{a_i} \tag{3.5}$$

with $c = (c_1, c_2, \cdots, c_n)$. Set $a = \sum_{i=1}^{n} a_i$, we obtain the function $c \mapsto c^{\frac{1}{a}}$, which is strictly monotone for $c > 0$.

$$u(c) = \tilde{u}(c)^{\frac{1}{a}} \tag{3.6}$$

represents the same preferences. Setting $\rho_i = a_i/a$ it can be shown that

$$u(c) = \prod_{i=1}^{n} c_i^{\rho_i}, \qquad \sum_{i=1}^{n} \rho_i = 1 \tag{3.7}$$

The problem of maximum utility is solved by looking at the logarithm of the utility

$$\max_{c} \sum_{i=1}^{n} \rho_i \ln c_i \tag{3.8}$$

3.2.3   Strategy choice and Nash equilibrium

The Nash equilibrium is a certain combination of strategy choices, under which no player can benefit by unilaterally changing his strategy while those of the other players remain unchanged. The Nash equilibrium is based on the assumption that all players are rational and that their rationality is common knowledge.

A formal definition of the Nash equilibrium is as follows. Let $G = <P, S, U>$ be a game and s_i be a strategy profile of all players except for player i. After each player i has chosen their strategies, player i obtains payoff $u_i(s_1, \cdots, s_n)$. Note that the payoff depends on the strategy chosen by player i as well as the strategies chosen by all the other players. A strategy profile $\{s_1^*, \cdots, s_n^*\} \in S$ is a Nash equilibrium if no unilateral deviation in strategy by any single player is profitable for that player, that is

$$\forall i, s_i \in S_i, s_i \neq s_i^* : u_i(s_i^*, s_{-i}^*) > u_i(s_i, s_{-i}^*) \tag{3.9}$$

Nash equilibrium analyzes a strategy profile under the assumption of complete information. However, if some information is private, and not known to all players, the players with incomplete information have to evaluate the possible strategy profiles. In particular, every rational player tries to take an action which maximizes its own expected payoffs, supposing a particular probability distribution of actions taken by other competitors. Therefore, the belief about which strategies other players will choose is crucial. Players can only make the best responses based on a correct belief. Each strategy is the best response to all other strategies in the Bayesian Nash equilibrium.

In Bayesian games, a type space T_i of player i is introduced, and each T_i has a probability distribution D_i. Assume that all players know $D_1, \cdots, D_n$, and the type t_i of player i is the outcome drawn from D_i independently.

A Bayesian Nash equilibrium is defined as a strategy profile with which every type of player is maximizing their expected payoffs given other type-contingent strategies. Especially for player i with the strategy $s_i : T_i \rightarrow S_i$, a strategy profile $\{s_1^*, \cdots, s_n^*\} \in S$ is a Bayesian Nash equilibrium if

$$\forall i, t_i \in T_i, s_i \in S_i, s_i \neq s_i^* : \tag{3.10}$$
$$E_{D_{-i}}[u_i(t_i, s_i^*(t_i), s_{-i}^*(t_{-i}))] > E_{D_{-i}}[u_i(t_i, s_i(t_i), s_{-i}^*(t_{-i}))]$$

However, Nash equilibrium may not be Pareto optimal from the global view. Nash equilibrium checks whether a profitable payoff exists when other payoffs are unchanged. Pareto efficiency examines whether a profitable payoff exists without reducing others's payoffs. Therefore, for the egocentric agents in the cloud market, the Nash equilibrium is more suitable than the Pareto efficiency to evaluate the allocation decisions.

3.3   Cloud resource allocation model

Virtualization technology hides heterogeneous configuration details from customers, and makes computation services functionally identical. Cloud users only need to choose a proper computing capacity that meets their requirements and pay according to the amount of usage. Cloud suppliers offer their customers more than one payment solution. For example, Amazon EC2 provides three different purchasing options: the on-demand model, the reserved model, and the spot model. Each model has different applicable scopes and limitations [Yi et al., 2010]. In order to satisfy more specific demands, we study a bid-based model as a complementary payment option to give users the flexibility to optimize their costs.

3.3.1   Bid-shared auction

In a cloud market, there are N users asking for services, each having a sequence of tasks to complete. The maximum number of tasks is K. A Cloud provider entirely virtualizes K resources, each of which can render a specific service with a fixed finite capacity C.

$$\mathbf{C} = [C_1, C_2, \ldots, C_K] \tag{3.11}$$

We characterize one task by its size, which means the amount of computing capability required to complete the task.

$$\mathbf{q} = \begin{bmatrix} q_1^1 & \cdots & q_k^1 & \cdots & q_K^1 \\ \vdots & \ddots & \vdots & \ddots & \vdots \\ q_1^i & \cdots & q_k^i & \cdots & q_K^i \\ \vdots & \ddots & \vdots & \ddots & \vdots \\ q_1^N & \cdots & q_k^N & \cdots & q_K^N \end{bmatrix} \tag{3.12}$$

Not all users have the same task itinerary; the size of an inexistent task is zero in the above matrix $\mathbf{q}$. If a task q_k^i can occupy its corresponding resource C_k, the computation is processed fastest, at a speed of $\omega_k^i = q_k^i/C_k$. However, in our model, resource capacity is never for exclusive use but shared by multiple users. It is reasonable and fair that resource partition is proportional to the user's outlay. We assume that a resource is always fully utilized and unaffected by how it is partitioned among users.

In the real commodity market, consumers needing the same commodity are competitors, and are reluctant to cooperate with each other. Thus, resource allocation in clouds is a non-cooperative allocation problem.

Every user has a bidding function, which decides the bid in any round considering task size, priority, QoS requirement, budget and deadline. The repeated bidding behavior is considered as a stochastic process indexed by a discrete time set. The outputs are random variables that have certain distributions, when these above deterministic arguments and time are fixed.

$$\{B^i(k), k \in (1, 2, \ldots, K)\} \tag{3.13}$$

Where B^i is the money that a user is willing to pay for one unit of resource per second. User i bids for task k at price b_k^i, which can be treated as a sample for B^i.

$$\mathbf{B} = \begin{bmatrix} B^1 \\ \vdots \\ B^i \\ \vdots \\ B^N \end{bmatrix} = \begin{bmatrix} b_1^1 & \cdots & b_k^1 & \cdots & b_K^1 \\ \vdots & \ddots & \vdots & \ddots & \vdots \\ b_1^i & \cdots & b_k^i & \cdots & b_K^i \\ \vdots & \ddots & \vdots & \ddots & \vdots \\ b_1^N & \cdots & b_k^N & \cdots & b_K^N \end{bmatrix} \tag{3.14}$$

The sum Θ_k of total bids for task k indicates the resource price.

$$\Theta_k = \sum_{i=1}^{N} b_k^i \tag{3.15}$$

Meanwhile, $\theta_k^{-i} = \sum_{j \neq i}^{N} b_k^j$ is given as the sum of other bids except bid b_k^i.

The bid-shared model indicates that resource k obtained by the user i is proportional to their bid price. The portion is $x_k^i = \frac{b_k^i}{\sum_{i=1}^{N} b_k^i}$, and obviously,

$$\forall k, \sum_{i=1}^{N} x_k^i = 1$$

Time spent on task k is defined by

$$t_k^i = \frac{q_k^i}{C_k x_k^i} = \omega_k^i + \omega_k^i \frac{\theta_k^{-i}}{b_k^i} \tag{3.16}$$

Cost taken to complete task k is

$$e_k^i = b_k^i t_k^i = \omega_k^i \theta_k^{-i} + \omega_k^i b_k^i \tag{3.17}$$

Two illuminations are obtained from the time and cost functions.

3.3.2 Non-cooperative game

Both time and expenditure depend not only on b_k^i that a user is willing to pay, but also on θ_k^{-i} that other competitors will pay. We therefore construct a non-cooperative game to analyze the bid-shared model.

In games, the set of players is denoted by N cloud users. Any player i independently chooses the strategy b_k^i from their strategy space B^i. The preference is determined by payoff, for example, we take computation time t_k^i as the payoff. Each player wishes their tasks to be computed as fast as possible, so the lower the payoff value is, the better. Regardless of the value of θ_k^{-i}, the dominant strategy of player i

is a low value of b_k^i if they want to get the optimal payoff. On the contrary, when we choose cost as the game payoff, the dominated strategy is high value of b_k^i, which is different from the former dominated strategy. This difference alerts us that the payoff must be carefully selected in order to indicate the outcome preference of a game. Absolute dependence on time or money is unreasonable.

We combine cost expense and computation time into an aggregate quantity, which stands for the total amount of substituted consumption. Similar to the utility function discussed above, constant elasticity of the substitution function indicates the players' payoff

$$C = \frac{\rho_e \ln \sum_{k=1}^{K} e_k^i + \rho_t \ln \sum_{k=1}^{K} t_k^i}{\rho_e + \rho_t} \tag{3.18}$$

where ρ_e, ρ_t are the output elasticities of cost and time, respectively.

3.4   Nash equilibrium allocation algorithms

3.4.1   Bid functions

In a cloud market, customers are rational decision makers who seek to minimize their consumption, and have constraints of cost $E = [E^1, E^2, \dots, E^N]$ and time $T = T[T^1, T^2, \dots, T^N]$. With a limited budget E^i and deadline T^i, the optimal object function of user i is

$$\text{Min} \quad C$$

$$\text{s.t.} \sum_{k=1}^{K} e_k^i \leq E^i \tag{3.19}$$
$$\sum_{k=1}^{K} t_k^i \leq T^i$$

The Hamilton equation is built by introducing the Lagrangian

$$\mathcal{L} = \frac{\rho_e \ln \sum_{k=1}^{K} e_k^i + \rho_t \ln \sum_{k=1}^{K} t_k^i}{\rho_e + \rho_t}$$

$$+ \lambda_e^i (\sum_{k=1}^{K} e_k^i - E^i) + \lambda_t^i (\sum_{k=1}^{K} t_k^i - T^i)$$

$$= \frac{\rho_e \ln \sum_{k=1}^{K} (\omega_k^i \theta_k^{-i} + \omega_k^i b_k^i) + \rho_t \ln \sum_{k=1}^{K} (\omega_k^i + \omega_k^i \frac{\theta_k^{-i}}{b_k^i})}{\rho_e + \rho_t}$$

$$+ \lambda_e^i (\sum_{k=1}^{K} (\omega_k^i \theta_k^{-i} + \omega_k^i b_k^i) - E^i) + \lambda_t^i (\sum_{k=1}^{K} (\omega_k^i + \omega_k^i \frac{\theta_k^{-i}}{b_k^i}) - T^i)$$

$\mathcal{L}$ is a function of three variables of b_k^i, λ_e^i and λ_t^i. To obtain the dynamic extreme point, the gradient vector is set to zero.

$$\nabla \mathcal{L}(b_k^i, \lambda_e^i, \lambda_t^i) = 0 \tag{3.20}$$

1. Take partial derivative with respect to b_k^i

$$\frac{\partial \mathcal{L}}{\partial b_k^i} = \frac{\rho_e}{\rho_e + \rho_t} \frac{\omega_k^i}{\sum e_k^i} - \frac{\rho_t}{\rho_e + \rho_t} \frac{\omega_k^i \theta_k^{-i}}{\sum t_k^i {b_k^i}^2} + \lambda_e^i \omega_k^i - \lambda_t^i \frac{\omega_k^i \theta_k^{-i}}{{b_k^i}^2} = 0 \quad (3.21)$$

which gives

$$\frac{\frac{\rho_e}{\sum e_k^i} + \lambda_e^i}{\frac{\rho_t}{\sum t_k^i} + \lambda_t^i} = \frac{\theta_k^{-i}}{(b_k^i)^2} \quad (3.22)$$

A similar result is obtained by setting the gradient of $\mathcal{L}$ at b_j^i to zero $\frac{\partial \mathcal{L}}{\partial b_j^i} = 0$,

$$\frac{\frac{\rho_e}{\sum e_k^i} + \lambda_e^i}{\frac{\rho_t}{\sum t_k^i} + \lambda_t^i} = \frac{\theta_j^{-i}}{(b_j^i)^2} \quad (3.23)$$

For user i, the capital sum $\sum e_k^i$ and time sum $\sum t_k^i$ remain the same for any two tasks; we could therefore determine the relationship between any two bids in one task sequence, which is

$$\frac{\theta_k^{-i}}{(b_k^i)^2} = \frac{\theta_j^{-i}}{(b_j^i)^2} \quad (3.24)$$

Then bid k is expressed by bid j, $b_k^i = b_j^i \sqrt{\frac{\theta_k^{-i}}{\theta_j^{-i}}}$.

Given Θ_k, preferences ρ_e and ρ_t exert major influence on bids. To be more specific, $\rho_e > \rho_t$ reveals that one user prefers satisfying budget to deadline, otherwise, the deadline constraint is more important than cost consumption.

2. Take partial derivative with respect to λ_e^i

$$\frac{\partial \mathcal{L}}{\partial \lambda_e^i} = \sum_{k=1}^{K} e_k^i - E^i = \sum_{k=1}^{K} \omega_k^i (b_k^i + \theta_k^{-i}) - E^i = 0 \quad (3.25)$$

Substituting b_j^i for $\sqrt{\frac{\theta_j^{-i}}{\theta_k^{-i}}} b_k^i$, the equation is expanded

$$\sum_{j=1}^{k-1} \omega_j^i \left(\sqrt{\frac{\theta_j^{-i}}{\theta_k^{-i}}} b_k^i + \theta_j^{-i} \right) + \omega_k^i (b_k^i + \theta_k^{-i})$$
$$+ \sum_{j=k+1}^{K} \omega_j^i \left(\sqrt{\frac{\theta_j^{\hat{-}i}}{\theta_k^{-i}}} b_k^i + \theta_j^{\hat{-}i} \right) - E^i = 0 \quad (3.26)$$

Simplifying the above equation, user i will bid for task k at price

$$b_k^i = \frac{E^i - \sum_{j=1}^{k-1} \omega_j^i \theta_j^{-i} - \omega_k^i \theta_k^{-i} - \sum_{j=k+1}^{K} \omega_j^i \theta_j^{\hat{-}i}}{\sum_{j=1}^{k-1} \omega_j^i \sqrt{\frac{\theta_j^{-i}}{\theta_k^{-i}}} + \omega_k^i + \sum_{j=k+1}^{K} \omega_j^i \sqrt{\frac{\theta_j^{\hat{-}i}}{\theta_k^{-i}}}} \quad (3.27)$$

3. Take partial derivative with respect to λ_t^i

$$\frac{\partial \mathcal{L}}{\partial \lambda_t^i} = \sum_{k=1}^{K} t_k^i - T^i = \sum_{k=1}^{K} \frac{\omega_k^i (b_k^i + \theta_k^{-i})}{b_k^i} - T^i = 0 \qquad (3.28)$$

The expanded expression is obtained

$$\sum_{j=1}^{k-1} \omega_j^i \left(\frac{\sqrt{\frac{\theta_j^{-i}}{\theta_k^{-i}}} b_k^i + \theta_j^{-i}}{\sqrt{\frac{\theta_j^{-i}}{\theta_k^{-i}}} b_k^i} \right) + \omega_k^i \left(\frac{b_k^i + \theta_k^{-i}}{b_k^i} \right) + \sum_{j=k+1}^{K} \omega_j^i \left(\frac{\sqrt{\frac{\hat{\theta}_j^{-i}}{\theta_k^{-i}}} b_k^i + \hat{\theta}_j^{-i}}{\sqrt{\frac{\hat{\theta}_j^{-i}}{\theta_k^{-i}}} b_k^i} \right) - T^i$$

$$= 0 \qquad (3.29)$$

The above equation is further simplified by

$$b_k^i = \frac{\sum_{j=1}^{k-1} \omega_j^i \sqrt{\theta_j^{-i} \theta_k^{-i}} + \omega_k^i \theta_k^{-i} + \sum_{j=k+1}^{K} \omega_j^i \sqrt{\hat{\theta}_j^{-i} \theta_k^{-i}}}{T^i - \sum_{j=1}^{K} \omega_j^i} \qquad (3.30)$$

Equation (3.27) and equation (3.30) show the influences of budget and deadline on bidding price b_k^i, respectively. Both equations reveal that current bid b_k^i is decided by competitors' bids in past $\theta_j^{-i}(j < k)$, present θ_k^{-i}, and future $\theta_j^{-i}(j > k)$. If bidding functions are based on the assumption that all other payments are fixed throughout the network, the model is classified as a static game of complete information [Gibbons, 1992]. However, these isolated cloud users are unable to collect all rivals' financial information in a real market, and the resource allocation problem evolves into the game of incomplete information. In that case, b_k^i is a function with respect to a vector $[\theta_1^{-i}, \cdots, \theta_k^{-i}, \hat{\theta}_{k+1}^{-i}, \cdots, \hat{\theta}_K^{-i}]$, only if the expectation of future bids $\hat{\theta}_{k+1}^{-i}, \cdots, \hat{\theta}_K^{-i}$ are estimated precisely.

3.4.2   Parameters estimation

The existence of the Nash Equilibrium with complete information has been proved by Bredin [Bredin et al., 2003]. However, new problems arise when buyers do not intend to expose their bids to other competitors or when they are allowed to join or leave a datacenter from time to time. How does one deal with the lack of information? How do users predict the price trend on the basis of inadequate knowledge? We record historical purchasing prices $\Theta_1, \cdots \Theta_{k-1}$ in past auctions, and then use statistical forecasting method to evaluate the future price.

In probability theory, Bayes' theorem shows how the probability of a hypothesis depends on its inverse if observed evidence is given. The *posteriori* distribution can be calculated from the *priori* $p(\Theta)$, and its likelihood function $p(\Theta \mid \Theta_k)$ is

$$p(\Theta \mid \Theta_k) = \frac{p(\Theta_k|\Theta)p(\Theta)}{\int p(\Theta_k|\Theta)p(\Theta)d\Theta} \qquad (3.31)$$

The *posteriori* hyperparameters $p(\Theta|\Theta_k)$ can be achieved by using the Bayesian learning mechanism, the value of which determines the maximum likelihood prediction of resource price. So future bids are forecasted as

$$\hat{\theta}_{k+1}^{-i} = \mathrm{E}(\Theta|\Theta_k) - \mathrm{E}(B^i)$$
$$\vdots \qquad\qquad\qquad\qquad (3.32)$$
$$\hat{\theta}_K^{-i} = \mathrm{E}(\Theta|\Theta_{K-1}) - \mathrm{E}(B^i)$$

Three parameters α_k^i, β_k^i and γ_k^i are introduced, which stand for information from other competitors.

$$\alpha_k^i = \sum_{j=1}^{k-1} \omega_j^i \theta_j^{-i} + \sum_{j=k+1}^{K} \omega_j^i \hat{\theta}_j^{-i}$$
$$\beta_k^i = \sum_{j=1}^{k-1} \omega_j^i \sqrt{\theta_j^{-i}} + \sum_{j=k+1}^{K} \omega_j^i \sqrt{\hat{\theta}_j^{-i}} \qquad (3.33)$$
$$\gamma_k^i = \sum_{j=1}^{k-1} \omega_j^i + \sum_{j=k+1}^{K} \omega_j^i$$

Substituting θ_k^{-i} by $\Theta_k - b_k^i$ in equation (3.27), we obtain the explicit function $f_k^i(\Theta_k)$ with respect of Θ_k.

$$f_k^i(\Theta_k) = \frac{(E^i - \alpha_k^i - \omega_k^i \Theta_k)^2}{2(\beta_k^i)^2} \left(\sqrt{1 + \frac{4(\beta_k^i)^2 \Theta_k}{(E^i - \alpha_k^i - \omega_k^i \Theta_k)^2}} - 1 \right) \qquad (3.34)$$

Figure 3.2 shows that task bid is decided not only by its budget, but also by its

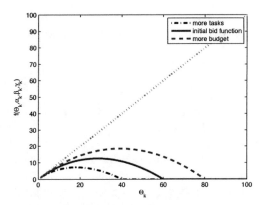

FIGURE 3.2: Bid under budget constraint.

workloads. Compared with the solid line, the dot dash line shows that a wealthier user is capable of submitting a larger positive bid and has a larger participated bid range. On the contrary, the user with a heavy workload has to save cash for the following competitions, so the money allocated to the current task is very limited, which is shown by softened dash line.

Substituting θ_k^{-i} by $\Theta_k - b_k^i$ in equation (3.30), we obtain the explicit function $g_k^i(\Theta_k)$ with respect of Θ_k, which characterizes bid price under deadline constraint.

$$g_k^i(\Theta_k) = \frac{\omega_k^i}{T^i - \gamma_k^i}\Theta_k + \frac{\sqrt{(\beta_k^i)^4 + 4(\beta_k^i)^2(T^i - \gamma_k^i)(T^i - \gamma_k^i - \omega_k^i)\Theta_k}}{2(T^i - \gamma_k^i)^2} - \frac{(\beta_k^i)^2}{2(T^i - \gamma_k^i)^2} \quad (3.35)$$

As seen from equation (3.35), $g_k^i(\Theta_k)$ is a monotone increasing function with respect to Θ_k, which means that bids can grow to infinite if the budget constraint is omitted. Obviously, exorbitant price would not deter the users who have sufficient capital, so vicious competition can not be restrained. In Figure 3.3, the dot dash line illustrates

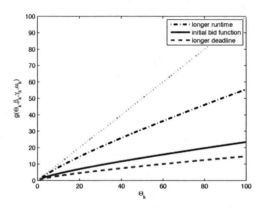

FIGURE 3.3: Bid under deadline constraint.

that one user will not be in a hurry to make a high bid for sufficient resource if he has enough time. Thus, the user can control his expenditure more effectively. As seen from softened dash line, longer task runtime needs more computing capacity, so the bidding price rises accordingly. The bid functions under budget and deadline constraints are compared in Figure 3.4. The range of possible bids enlarges accordingly when constraints are loosened. The intersection of the two solid lines signifies that budget and deadline are both exhausted at the same time. If deadline is extended, the solid budget curve meets the dashed deadline curve at a lower position. It indicates that the possible bid should be above the solid deadline curve in order to complete all tasks in finite time. For the same reason, if one user holds more funds, the intersection moves right along the solid deadline curve, so the left side of solid budget curve will contain the possible bids. The bid region is surrounded by cross and plus curves. Specifically, the crosses mean that all capital is used up with time remaining, while the pluses mean that deadline is reached with redundant money. Outside this region, there is no feasible bidding solution, which indicates the given constraints are

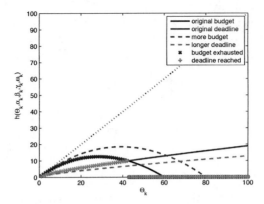

FIGURE 3.4: Bid under double constraints.

too rigid. Users must loosen either of the two constraints slightly if they still wish to accomplish this impossible mission. Furthermore, regardless of whether the budget or deadline constraints are relaxed, the range Θ_k over which users can participate is stretched.

The cross curve is chosen as the new bidding function $h_k^i(\Theta_k)$ under double constraints, because higher bids are more competitive in terms of a fixed Θ_k.

$$h_k^i(\Theta_k) = \begin{cases} f_k^i(\Theta_k) & : \quad f_k^i(\Theta_k) \geq g_k^i(\Theta_k) \\ 0 & : \quad f_k^i(\Theta_k) < g_k^i(\Theta_k) \end{cases} \tag{3.36}$$

3.4.3 Equilibrium price

The bid functions of any user i have been deduced. Next, we analyze whether an equilibrium price exists and how it is obtained.

In the beginning, users who need resource k make their initial bids,

$$\Theta_k^{(1)} = \sum_N b_k^i \tag{3.37}$$

In the first round, money that users are asked to pay for the resource partition is calculated by bid function $h_k^i(\Theta_k^{(1)})$. A general expression is

$$b_k^{i(m)} = h_k^i(\Theta_k^{(m)}) \tag{3.38}$$

where m means values are in the mth round. Hence, the price that the cloud provider prepares to charge from N users is actually

$$\Theta_k^{(m+1)} = \sum_N h_k^i(\Theta_k^{(m)}) \tag{3.39}$$

The corresponding partition is $x_k^{i(m)} = b_k^{i(m)}/\Theta_k^{(m+1)}$. If anyone disagrees with the allocation due to either insufficient resource, or high cost, iteration will continue. Users can adjust their bids in the next round. If all users satisfy their allocation proportions, the current price

$$\Theta_k^{(m+1)} = \Theta_k^{(m)} \tag{3.40}$$

The resource price $\Theta_k^{(m+1)}$ is agreed to by every user, so this is an equilibrium price.

In game theory, the Nash Equilibrium occurs when no user can obtain more resource by changing his bid while others keep theirs unchanged, that is

$$b_k^{i*} = \text{Max} \quad x(b_k^i, \theta_k^{-i*}) \tag{3.41}$$

Where b_k^{i*} is equilibrium bid and $\theta_k^{-i*} = \Theta_k^* - b_k^{i*}$ is equilibrium performance of his competitors. When demand is higher than provision $\sum_N x_k^i > 1$, users tend to pay more to improve their own allocation proportion, so the resource price increases accordingly. High resource price will then reduce x_k^i until $\sum_N x_k^i$ approaches one. The reverse situation $\sum_N x_k^i < 1$ is also true. In conclusion, resource price has a negative impact on the value of $\sum_N x_k^i$, and pushes it to the situation where the resource is fully utilized $\sum_N x_k^{i(m)} = 1$. Therefore, $\sum_N x_k^i$ can be considered as a descending function with respect of Θ_k. Different resource prices $\Theta_{k1}^* \neq \Theta_{k2}^*$ have different values of $\sum_N x_k^i$, so the equilibrium price Θ_k^* that let $\sum_N x_k^i = 1$ is unique and a Nash Equilibrium exists. Figure 3.5 shows the equilibrium resource price for a

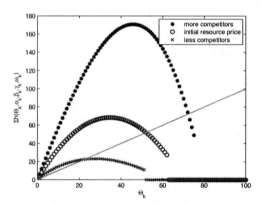

FIGURE 3.5: Equilibrium resource price under double constraints.

dynamic system under the condition that all users have similar bid distributions. The line with a slope equal to one shows that the bid function sum $\sum h(\Theta_k)$ of task k is equal to Θ_k. The intersection of this line and the curve $\sum h(\Theta_k)$ stands for the only stable solution. From this figure, we can observe how the final equilibrium price is affected by different numbers of users. An increasing number of competitors raises the bid sum and makes the resource more expensive. A user has to bid against more competitors if he really needs this resource. As a result, the resource price soars. Once the price becomes too high, some users quit the competitive bidding and the resource price will consequently decrease quickly.

3.5   Implementation in a cloud datacenter

Although there are several commercial cloud computing infrastructures, such as Aneka, Azure, EC2 and Google App Engine, building a cloud testbed on a real infrastructure is expensive and time consuming. It is impossible to evaluate performances of various application scenarios in a repeatable and controllable manner. We therefore apply simulation methodology for performance evaluation of resource allocation algorithms.

3.5.1   Cloudsim toolkit

Cloudsim [Buyya et al., 2009b] is designed to emulate cloud-based infrastructure and application service, and can be used in research on economy driven resource management policies on large scale cloud computing systems. Researchers benefit from focusing on resource allocation problems without implementation details. These features are not supported by other cloud simulators [Buyya et al., 2009b].

We apply Cloudsim as our simulation framework, but make some improvements aiming at the following shortcomings. First, sequential auctions are complemented, accompanied by several specific policies. Second, Cloudsim only supports static assignment with pre-determined resources and tasks. We realize that multiple users can submit their tasks over time according to certain arrival rates or probability distribution and that resource nodes can freely join or leave a cloud datacenter. The assignment in our simulation model is much closer to a real market than before.

3.5.2   Communication among entities

There are four types of entities to be simulated. CIS Registry provides a database level match-making service for mapping application requests to a datacenter. A datacenter integrates distributed hardware, database, storage devices, application software and operating systems to build a resource pool, and is in charge of virtualizing applicable computing resources according to users' requests. Cloud users have independent task sequences, and they purchase resources from datacenters to execute tasks. All these users bid according to their economic capabilities and priorities under different constraints. An auctioneer is the middleman in charge of maintaining an open, fair and equitable market environment. In accordance with the rules of market economy, the auctioneer fixes an equilibrium price for non-cooperative users to avoid blind competition. Figure 3.6 depicts the flow of communication among

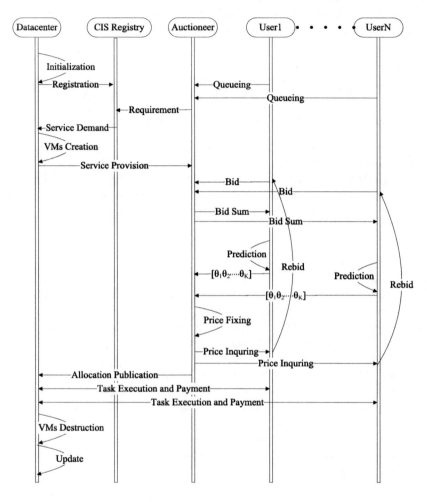

FIGURE 3.6: Flowchart of communication among entities.

the main entities. At the beginning, the datacenter initializes current available hosts, generating provision information and registers in CIS. Meanwhile, cloud users who have new tasks report to the auctioneer and queue up in order of arrival time. At regular intervals, the auctioneer collects information and requests the datacenter to virtualize corresponding resources. Once virtual machines are ready according to users' service requirements, the datacenter sends the provision information to the auctioneer, and successive auctions start.

In each auction stage, users ask the auctioneer individually about configuration information such as virtual machine provision policy, time zone, bandwidth, residual computing processors, and then bid according to their asset valuations. The auctioneer collects all bids, then informs users of the sum of bids. Under the game of incomplete information, cloud users only know their own price functions as well as the incurred sum of bids. They dynamically predicate the future resource price, and update competitors' information $[\theta_1^{-i}, \cdots, \theta_k^{-i}, \hat{\theta}_{k+1}^{-i}, \cdots, \theta_K^{-i}]$. Subsequently, holding all price functions, the auctioneer makes an equilibrium allocation decision and inquires whether everyone is satisfied with the result. If the result is agreeable, the auctioneer publishes allocation proportions to the datacenter and users. Users then execute their tasks and pay for the resource allocated. At the end, the datacenter deletes the used VMs and waits for new service demands.

3.5.3 Bidding algorithms

Concrete algorithms for users and auctioneer are explained in more detail by Algorithm 3.5.1 and Algorithm 3.5.2.

From a user's point of view, after task submission, an observer focuses on analyzing the received messages that prescribe the user's next move. If the auctioneer announces a new auction, the user adds it to the auction list. If bids are called, an appropriate bid is calculated and reported to the auctioneer. If the user receives the message calling for parameters, he examines the historical prices and estimates the future bid sum by Bayesian learning mechanism, then sends information back. Finally, if the user receives the resource price and proportion, he immediately updates his price list and begins to execute the task. From an auctioneer's perspective, a new auction is triggered whenever a new type of task arrives. Once an auction begins, the auctioneer broadcasts the bid calling message to current users. As soon as all proposals arrive, the auctioneer informs users of the sum Θ_k. Similarly, the auctioneer collects bidding function parameters from all the bidders, and then decides a reasonable bound. If the bound is too narrow, poor users quit gambling. The resource price is modified repeatedly until the difference between $\sum h_k^i$ and Θ_k is less than a predetermined threshold. Once the equilibrium price is found, allocation proportions are broadcast to all cloud users. After that, the auctioneer deletes the current auction and waits for a new task request.

3.5.4 Comparison of forecasting methods

First, normal distribution is used to describe the financial capability of the users. Bidding function B^i has mean μ^i and variance σ^2. We choose one user as our observable object, and assign a mean purchasing price of 10\$/s and a bid variance of 0.1. Other mean bids are generated randomly in the range of 1–100\$/s. This user is unaware of other economic situations, but keeps on estimating others from their prior behaviors. Figure 3.7 illustrates how the closing price changes as time goes by. We conclude that budget exerts a huge influence on preliminary equilibrium price,

---

Algorithm 3.5.1 User i bidding algorithm.

1: submit tasks to auctioneer
2: **if** observer receives message of inform start **then**
3:     add current auction
4: **end if**
5: **if** observer receives message of call for bids **then**
6:     set $\{b_1^i, \cdots, b_{k-1}^i\} \leftarrow b_k^i$
7:     send message of proposal to auctioneer
8: **end if**
9: **if** observer receives message of call for parameters **then**
10:     inquiry historical price $\theta_1^{-i}, \cdots, \theta_k^{-i}$
11:     forecast future price $\hat{\theta}_{k+1}^{-i}, \cdots, \hat{\theta}_K^{-i}$
12:     send message of competitors information to auctioneer
13: **end if**
14: **if** observer receives message of resource price **then**
15:     $\{\Theta_1, \cdots, \Theta_{k-1}\} \leftarrow \Theta_k$
16:     send message of task execution to resource
17:     delete current auction
18: **end if**

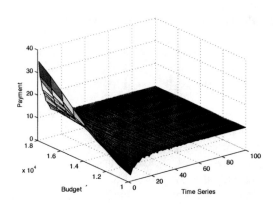

FIGURE 3.7: Convergence of Nash equilibrium bid.

because self-interested but rational users always wish to seek extra benefits from others. With limited budget, the user will behave conservatively at the initial stages, to avoid overrunning the budget and to save enough money to complete remaining tasks. Therefore, in the beginning, the equilibrium price is lower than the mean price. On the contrary, if the user has sufficient capital, he is eager to improve current payment to get a larger proportion. Competition leads the equilibrium price to rise, higher than the anticipated cost. However, with the money available for the current job decreasing, the user becomes less aggressive. As bidding is underway, price

---

Algorithm 3.5.2 Auctioneer allocation algorithm.

Require: $N \geq 2$

  1: initialize auctioneer
  2: **while** auction k **do**
  3:     set bidders to auction k
  4:     broadcast message to call for bids
  5:     **while** bidder's proposal arrives **do**
  6:         collect proposal message from bidder
  7:     **end while**
  8:     broadcast message to inform Θ_k
  9:     **while** bidder's parameter arrives **do**
 10:         collect parameter message from bidder
 11:     **end while**
 12:     **while** bidders disagree proportion **do**
 13:         **for all** cloud users **do**
 14:             build new bid function h_k^i
 15:         **end for**
 16:         $difference = \sum h_k^i - \Theta_k$
 17:         **if** $difference > threshold$ **then**
 18:             $\Theta_k = \sum h_k^i$
 19:         **else**
 20:             exit
 21:         **end if**
 22:         update vector $[\theta_1^{-i}, \cdots, \theta_k^{-i}, \hat{\theta}_{k+1}^{-i}, \cdots, \hat{\theta}_K^{-i}]$
 23:     **end while**
 24:     broadcast message to inform resource price
 25:     stop the current and wait for a new auction
 26: **end while**
 27: delete auctioneer

---

will gradually converge to the original mean value. Next, the accuracy of Bayesian learning prediction is evaluated when the cloud market is full of uncertainties, such as insufficient common knowledge and on-line task submitting. Figure 3.8 exhibits the predication of resource price in a dynamic game of incomplete information. If the common knowledge is insufficient, the user experientially predicts other bids using the published equilibrium prices. When the bidding variance is low, no more than 0.01, the estimation works quite well. Our policy differs a little from the scheme that hypothesizes that all users' information is fixed and public. If users perform unstably in the gambling process and the offered bids are more random, accurate price forecast becomes difficult. Provided that rivals' information is learned iteratively, experiment results show that the resource price still converges to the equilibrium price stage by stage.

Three forecasting methods are compared, including Bayesian learning, historical averaging and last-value following. Figure 3.9 shows the standard deviations of three

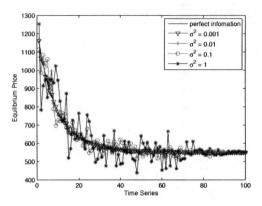

FIGURE 3.8: Prediction of resource price.

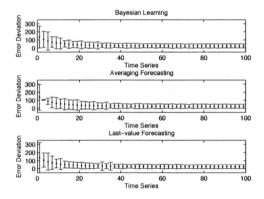

FIGURE 3.9: Forecast errors with normal distribution.

forecast methods versus time series. All three forecasting methods are able to converge to the result with perfect information, as long as the user keeps on training his estimates of others' bid functions over time. The cases with abundant budgets are examined. Some users would like to increase bids to obtain more resource, so the price keeps rising, to much higher than the estimated bid. If all the historical prices are used for prediction, the history averaging method behaves poorly at the beginning of auctions, and is less stable than the other two methods. Compared with the last-value method, Bayesian learning converges in a smoother manner, because historical prices are used to calculate the likelihood function, rather than simply following the price in the previous auction as in the last-value method. Now we apply another distribution, Pareto, to express users' bidding rules, meanwhile keeping other experiment setups the same. A similar conclusion can be reached in Figure 3.10, except that the worst forecast is the last-value method. The result is due to the attribute of

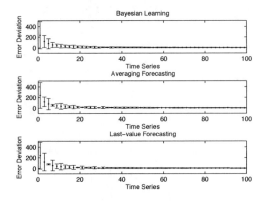

FIGURE 3.10: Forecast errors with Pareto distribution.

Pareto distribution. The Pareto principle stands for the probability that the variable is greater than its minimum, while normal distribution reveals how close data clusters are around its mean. For one specific round of bidding, it's more difficult to estimate the precise value with Pareto distribution than with normal distribution. In other words, the more historical data is accumulated, the more accurate the forecast will be. In Figure 3.10, convergence of Bayesian learning is still the most stable one of the three schemes. As a result, it is recommended as a forecast method in practical applications.

3.6 Concluding remarks

In this chapter, we solved the resource allocation problem in the user-level of cloud scheduling. We surveyed game theory, covering the different classes of games and their applications, payoff choice and utility function, as well as strategic choice and Nash equilibrium. Based on that, we built a non-cooperative game to solve the multi-user allocation problem in a cloud scenario. The scheduling model includes bid-shared auction, user strategy (bid function), price forecasting and equilibrium analysis.

We then proposed game theoretical algorithms for user bidding and auctioneer pricing. We supplemented a bid-shared auction scheme in a cloud simulation framework, Cloudsim, in order to realize sequential games. Results show that resource allocation can reach Nash equilibrium among non-cooperative users even though common knowledge is insufficient, and that the Bayesian learning forecast has the best and most stable performance. Therefore, our algorithms support financially smart customers with an effective forecasting method and help an auctioneer decide on an equilibrium resource price, so that they can potentially solve resource allocation problems in cloud computing.

Chapter 4

Multi-dimensional data analysis in a cloud datacenter

4.1 Introduction

Multi-dimensional data analysis applications are largely used in computing technologies for the identification, discovery and analysis of business data. Enterprises generate massive amounts of data every day. This data comes from various aspects of their products, for instance, the sale statistics of a series of products in each store. The raw data is extracted, transformed, cleansed and then stored under multi-dimensional data models, such as the star-schema[1]. Users query this data to help make business decisions. The queries are usually complex and involve large-scale data access. Here are some features for multi-dimensional data analysis queries:

- queries access large datasets performing read-intensive operations;

- queries are often quite complex and require different views of data;

- query processing involves many aggregations;

- updates can occur but infrequently, and can be planned by the administrator to execute at an expected time.

Data Warehouse is the type of software designed for multi-dimensional data analysis. However, faced with larger and larger volumes of data, the capacity of a centralized Data Warehouse seems too limited. The amount of data is increasing; the number of concurrent queries is also increasing. The scalability issue became a big challenge for centralized Data Warehouse software. In addition, short response time is also challenging for centralized Data Warehouse to process large-scale data. The solution addressing this challenge is to decompose and distribute the large-scaled dataset, and calculate the queries in parallel.

Three basic distributed hardware architectures exist, including shared-memory, shared-disk, and shared-nothing. Shared-memory and shared-disk architectures cannot scale well with the increasing dataset scale. The main reason is that these two distributed architectures require a large amount of data exchange over the interconnection network. However, the interconnection network cannot be infinitely expanded,

---

[1]A star schema consists of a single fact table and a set of dimension tables.

which becomes the main shortcoming of these two architectures. On the contrary, shared-nothing architecture minimizes resource sharing and therefore minimizes the resource contentions. It fully exploits local disk and memory provided by a commodity computer. It does not need a high-performance interconnection network, because it only exchanges small-sized messages over network. Such an approach minimizing network traffic allows more scalable design. Nowadays, the popular distributed systems have almost all adopted the shared-nothing architectures, including peer-to-peer, cluster, Grid, and the Cloud. The research work related to data over shared-nothing distributed architectures is also very rich. For instance, parallel database like Gamma [DeWitt et al., 1986], the DataGrid project [DataGrid Project, 2012], BigTable [Chang et al., 2008] etc. are all based on shared-nothing architecture.

In the distributed architecture, data is replicated on different nodes, and query is processed in parallel. To accelerate multi-dimensional data analysis query processing, many optimizing approaches were proposed. The traditional optimizing approaches, used in centralized Data Warehouse, mainly include pre-computing, indexing techniques and data partitioning. These approaches are still very useful in the distributed environment. In addition, a great deal of work has also been done to parallelize the query processing. In this chapter, we will talk about three approaches to accelerating query processing as well as their utilizations in a distributed environment. We will present parallelism of various operators, which are widely used in parallel query processing.

4.2   Pre-computing

The pre-computing approach resembles the materialized views optimizing mechanism used in database systems. In a multi-dimensional data context, the materialized views become **data cubes**[2]. Data cubes store the aggregates for all possible combinations of dimensions. These aggregates are used to answer the forthcoming queries.

For a cube with d attributes, the number of sub-cubes is 2^d. With the augmentation of the number of cube's dimensions, the total volume of a data cube will exponentially increase. Thus, such an approach produces data of a much larger volume than the original dataset, which might not have good scalability faced with the requirement of processing larger and larger datasets. Despite this, the pre-computing is still an efficient approach to accelerating query processing in a distributed environment.

---

[2]Data Cube is proposed in [Jim et al., 1998]; it is described as an operator, it is also called **cube** for short. Cube generalizes the histogram, cross-tabulations, roll-up, drill-down, and sub-total constructs, which are mostly calculated by data aggregation.

4.2.1   Data cube

Constructing a data cube in a distributed environment is one of the research topics. The reference [Sanjay and Alok, 1997] proposed some methods for the construction of data cubes on distributed-memory parallel computers. In their work, the data cube construction consists of six steps:

- Data partitioning among processors.

- Loading data into memory as a multi-dimensional array.

- Generating a schedule for the group-by aggregations.

- Performing the aggregation calculations.

- Redistributing the sub-cubes to processors for query processing.

- Defining local and distributed hierarchies on all dimensions.

In the data loading step (step 2), the size of the multi-dimensional array in each dimension equals the number of distinct values in each attribute; each record is represented as a cell indexed by the values[3] of each attribute. This step adopted two different methods for data loading: a hash-based method and a sort-based one. For small datasets, both methods work well, but for large datasets, the hash-based method works better than the sort-based method, because of its inefficient memory usage[4]. The cost for aggregating the measure values stored in each cell varies. The reason is that the whole data cube is partitioned over one or more dimensions, and thus some aggregating calculations involve the data located on other processors. For example, for a data cube consisting of three dimensions A, B and C, being partitioned over dimension A, then aggregations for the series of sub-cubes ABC→AB→A involve only local calculations on each node. However, the aggregations of sub-cube BC need the data from the other processors.

4.2.2   Sparse cube

Sparsity is an issue of data cube storage. In reality, the sparsity is a common case. Take an example of a data cube consisting of three dimensions (*product, store, customer*). If each store sells all products, then the aggregation over (product, store) produces $|product| \times |store|$ records. When the number of distinct values for each dimension increases, the product of the above formula will greatly exceed the number of records coming from the input relation table. When a customer enters a store, he/she cannot buy 5% of all the products. Thus, many records related to this customer will be an "empty cell." In the work of [Goil and Choudhary, 1999], the authors addressed the problem of sparsity in multi-dimensional array. In this work, data cubes

---

[3]The value of each attribute is a member of the distinct values of this attribute.
[4]The Sort-based method is accepted to work efficiently because in external memory algorithms it reduces the disk I/O over the hash-based method.

are divided into chunks, each chunk being a small equal-sized cube. All cells of a chunk are stored contiguously in memory. Some chunks, called *sparse chunks*, only contain sparse data. To compress the sparse chunks, they proposed a Bit-Encoded Sparse Storage (BESS) coding method. In this coding method, for a cell located in a sparse chunk, a dimension index is encoded in $\lceil \log |d_i| \rceil$ bits for each dimension d_i. They demonstrated that data compressed in this coding method could be used for efficient aggregation calculations.

4.2.3   Reuse of previous query results

Apart from utilizing pre-computed sub-cubes to accelerate query processing, one also tried to reuse the previous aggregate query results. With previous query results being cached in memory, if the next query, say Q_n, is evaluated to be contained within one of the previous queries, say Q_p, thus Q_n can be answered using the cached results calculated for Q_p. The case where Q_p and Q_n have an entire containment relationship is just a special case. In a more general case, the relationship between Q_p and Q_n is only overlap, which means only part of the cached results of Q_p can be used for Q_p. To address this partial-matching issue, the reference [Deshpande et al., 1998] proposed a chunk-based caching method to support fine granularity caching, allowing queries to partially reuse the results of previous queries with which they overlap. Another work [Liao and Pei, 2008] proposed a hybrid view caching method which gets the partially-matched result from the cache, and calculates the rest of the result from the component database, and then combines the cached data with calculated data to reach the final result.

4.2.4   Data compressing

Because the size of data cube growth is exponential with the number of dimensions, when the number of dimensions increases to a certain extent, the corresponding data cube will explode. In order to address this issue, some data cube compressing methods are proposed. For instance, Dwarf [Sismanis et al., 2002] is a method of constructing compressed data cubes. Dwarf considers eliminating prefix redundancy and suffix redundancy over cube computation and storage. The prefix redundancy commonly appears in the dense area, while the suffix redundancy appears in the sparse area. For a cube with dimensions (a, b, c), there are several group-bys, including a: (a, ab, ac, abc). Assuming that the dimension a has two distinct values a_1, a_2, dimension b has b_1, b_2 and dimension c has c_1, c_2, in the cells identified by (a_1, b_1, c_1), (a_1, b_1, c_2), (a_1, b_1), (a_1, b_2), (a_1, c_1), (a_1, c_2) and (a_1), the distinct value a_1 appears seven times, which causes a prefix redundancy. Dwarf can identify this kind of redundancy and store each unique prefix only once. For example, for aggregate values of three cells (a_1, b_1, c_1), (a_1, b_1, c_2), and (a_1, b_1), the prefix (a_1, b_1) is associated with one pointer pointing to a record with three elements (agg(a_1, b_1, c_1), agg(a_1, b_1, c_2), agg(a_1, b_1)). Thus the storage of cube cell identifiers, i.e. (a_1, b_1, c_1), (a_1, b_1, c_2), and (a_1, b_1), is reduced to storage of one prefix(a_1, b_1) and one pointer. The suffix redundancy occurs when two or more group-bys share a common suffix (like, (a, b, c) and (b, c)). If a and b are two correlated dimensions, some value of dimension a, say a_i, only appears together with another value b_j of dimension b. Then the cells (a_i, b_j, x) and (b_j, x) always have the same aggregate values. Such a suffix redundancy can be identified and eliminated by Dwarf during the construction of cube. Thus, for a cube of 25 dimensions of one petabyte, Dwarf reduces the space to 2.3 GB within 20 minutes.

The condensed cube [Wang et al., 2002] is also a way of reducing the size of a data cube. It reduces the size and the time required for computing the data cube. However, it does not adopt the approach of data compression. No data decompression is required to answer queries, nor is on-line aggregation required when processing queries. Thus, no additional cost is incurred during the query processing. The cube condensing scheme is based on the Base Single Tuple (**BST**) concept. Assume a base relation R $(A, B, C, \dots)$, and the data cube $Cube(A, B, C)$ is constructed from R. Assume that attribute A has a sequence of distinct values $a_1, a_2 \dots a_n$. Considering a certain distinct value a_k, where $1 \leq k \leq n$, if among all the records of R, there is only one record, say r, containing the distinct value a_k, then r is a BST over dimension A. To be noted, one record can be a BTS on more than one dimension. For example, continuing the previous description, if the record r contains distinct value c_j on attribute C, and no other record contains c_j on attribute C, then record r is the BST on C. The author gave a lemma saying that if record r is a BST for a set of dimensions, say SD, then r is also the BTS of the superset of SD. For example, consider record r, a BST on dimension A, then r is also the BST on (A, B), (A, C), (A, B, C). The set containing all these dimensions is called $SDSET$. The aggregates over the $SDSET$ of a same BST r always concern the same record r, which means that any aggregate function $aggr()$ only applies on record r. Thus, all these aggregate values will have equal value $aggr(r)$. Thus, only one unit of storage is required for the aggregates of the dimensions and combination of dimensions from $SDSET$.

4.3   Data indexing

Data indexing is an important database system technology, especially when good performance of read-intensive queries is critical. The index is composed of a set of particular data structures specially designed for optimizing the data access. When performing read operations over the raw dataset, within which the column values are randomly stored, only full table scan can achieve data item lookup. In contrast, when performing read operations over an index, where data items are specially organized, and the auxiliary data structures are added, the read operations can be performed much more efficiently. For queries of read-intensive characteristics, such as multi-dimensional data analysis query, index technology is an indispensable aid to accelerate query processing. Compared with other operations running within the memory, the operations for reading data from the disk might be the most costly. One real example cited from [O'Neil and Quass, 1997] can demonstrate this: "we assume 25 instructions needed to retrieve the proper records from each buffer resident page. Each disk page I/O requires several thousand instructions to perform." It is clear that data accessing operations are very expensive, especially if it becomes the most common operation in the read-intensive multi-dimensional data analysis application. Indexing data improves the data accessing efficiency by providing the particular data structures. The performance of index structures depends on different parameters, such as the number of stored records, the cardinality of the dataset, the disk page size of the system, the bandwidth of disks and latency time etc. The index techniques used in Data Warehouse come from the index of databases. Many useful indexing technologies are proposed, such as, B-tree/B^+-tree index [Douglas, 1979], projection index [O'Neil and Quass, 1997], Bitmap index [O'Neil and Quass, 1997], Bit-Sliced index [O'Neil and Quass, 1997], join index [Valduriez, 1987], inverted index [Cutting and Pedersen, 1990] etc. We will review these interesting index technologies in this section.

4.4   Data partitioning

In order to reduce the resource contention[5], a distributed parallel system often uses an affinity scheduling mechanism, giving each processor an affinity process to execute. Thus, in a shared-nothing architecture, this affinity mechanism tends to be realized by data partitioning; each processor processes only a certain fragment of the dataset. This forms the preliminary idea of data partitioning.

Data partitioning can be logical or physical. Physical data partitioning means reorganizing data into different partitions, while logical data partitioning will greatly affect physical partitioning. For example, a design used in Data Warehouse, namely data mart, is a subject-oriented, logical data partitioning. In a Data Warehouse built in an enterprise, each department is interested only in a part of the data. Then, the data partitioned and extracted from Data Warehouse for this department is referred to as a data mart. As we are more interested in the physical data access issue, we will focus on the physical data partitioning techniques.

4.4.1   Data partitioning methods

Data partitioning allows one to exploit the I/O bandwidth of multiple disks by reading and writing in parallel. That increases the I/O efficiency of disks without needing any specialized hardware [DeWitt and Gray, 1992]. Horizontal partitioning and vertical partitioning are two main methods of data partitioning.

---

[5]Resource contention includes disk bandwidth, memory, network bandwidth, etc.

4.4.1.1 Horizontal partitioning

Horizontal partitioning conserves the record's integrality. It divides tables, indexes and materialized views into disjoint sets of records that are stored and accessed separately. Previous studies show that horizontal partitioning is more suitable in the context of relational Data Warehouses [Bellatreche et al., 2004]. There are mainly three main types of horizontal partitioning, *round-robin partitioning*, *range partitioning* and *hash partitioning*.

Round-robin partitioning. Round-robin partitioning is the simplest strategy to dispatch records among partitions. Records are assigned to each partition in a round-robin fashion. Round-robin partitioning works well if the applications access all records in a sequential scan. Round-robin does not use a partitioning key, and then records are randomly dispatched to partitions. Another advantage of round-robin is that it provides good load balancing.

Range partitioning. Range partitioning uses a certain attribute as the partitioning attribute, and records are distributed among partitions according to the values of their partitioning attribute. Each partition contains a certain range of values on an indicated attribute. For example, table CUSTOMER_INFO stores information about all customers. We define column ZIP-CODE as the partition key. We can range-partition this table by giving zip-code as a rule between 75,000 and 75,019. The advantage of range partitioning is that it works well when applications sequentially or associatively access data[6], since records are clustered after being partitioned. Data clustering puts related data together in physical storage, i.e. the same disk pages. When applications read the related data, the disk I/Os are limited. Each time one disk page is read, not only the targeted data item, but also other needed data items of potential operations are fetched into memory. Thus, *Range partitioning* makes disk I/Os more efficient.

Hash partitioning. Hash partitioning is ideally suitable for applications that access data in a sequential manner. Hash partitioning also needs an attribute as the partitioning attribute. Records are assigned to a particular partition by applying a *hash* function over the partitioning key attribute of each record. Hash partitioning works well with both sequential data access applications and associative data access ones. It can also handle data with no particular order, such as alphanumeric product code keys.

The problem with horizontal partitioning is that it might cause data skew, where all required data for a query is put in one partition. Hash partitioning and round-robin partitioning are less likely to cause data skew, but range partitioning may cause this relatively easily.

---

[6]Associative data accessing means accessing all records holding a particular attribute value.

4.4.1.2   Vertical partitioning

Another data partitioning method is vertical partitioning, which divides the original table, index or materialized view into multiple partitions containing fewer columns. Each partition has a full number of records, but partial attributes. As each record has fewer attributes, the record size is smaller. Thus, each disk page can hold more records, which allows query processing to reduce disk I/Os. When the cardinality of the original table is large, this benefit is more obvious.

However, vertical partitioning has some disadvantages. First, updating (insert or delete) records in a vertically partitioned table involves operations over more processors. Second, vertical partitioning breaks the record integrity. Nevertheless, vertical partitioning is useful in some specific contexts. For example, it can separate frequently updated data (dynamic data) columns from static data columns; the dynamic data can be physically stored as a new table. In the processing of data read-intensive OLAP queries, vertical partitioning has its own advantages:

- By isolating certain columns, it is easier to access data, and create indexes over these columns [Datta et al., 2012, Datta et al., 1998].

- Column-specific Data compression, like run-length encoding, can be directly performed [Abadi et al., 2008].

- Multiple values from a column can be passed as a block from one operator to the next. If existing attributes have fixed-length values, then they can be iterated as an array [Abadi et al., 2008].

- For some specific datasets with a high dimension number, vertical partitioning is more reasonable in terms of time and space costs [Li et al., 2004].

4.4.2   Horizontal partitioning of a multi-dimensional dataset

The multi-dimensional dataset usually has a large volume. But the calculations over them are expected to run rapidly. As one of the OLAP query optimizing approaches, data partitioning makes it possible to process queries in a parallel and distributed fashion. Also, it can reduce irrelevant data accesses, improve the scalability, and ease data management. In this subsection, we summarize the applications of horizontal partitioning in OLAP query processing. By horizontal partitioning, we refer to the partitioning method that conserves the record integrity. Depending on the data-storage methods, OLAP tools can be categorized into Relational OLAP (ROLAP) and Multi-dimensional OLAP (MOLAP). In ROLAP, data is stored in the form of relations under star-schema. In MOLAP, data is stored in the form of cubes or multi-dimensional arrays, i.e., data cubes. We specify the data partitioning approaches designed for these two different data models separately.

4.4.2.1   Partitioning multi-dimensional array data

In a MOLAP, a data cube is represented as a multi-dimensional space, stored in optimized multi-dimensional array storage. In the multi-dimensional space, each *di-*

mension is represented as an axis; the distinct values of each dimension are various coordinate values on the corresponding axis. The *measures* are loaded from each record in the original dataset into the cells of this multi-dimensional space, each cell being indexed by the unique values of each attribute of the original record. Partitioning a data cube into dimensions and measures is a design choice [Sanjay and Alok, 1997].

Partitioning data cubes should support equal or near-equal distributions of work, i.e., the various aggregate computations for a set of cuboids, among processors. The partitioning approach should be dimension-aware, which means that it should provide some regularity in supporting dimension-oriented operations. Partitioning can be performed over one or more dimensions [Sanjay and Alok, 1997, Goil and Choudhary, 1999]. That is to say, the basic multi-dimensional array is partitioned on one or more dimensions. The dimensions over which the partitioning is performed are called partitioning dimensions. After partitioning, each processor holds a smaller multi-dimensional array, where the number of distinct values held in each partitioning dimension is smaller than in the whole multi-dimensional array. Thus, the distinct values over each partitioning dimension do not overlap among the sub-multi-dimensional arrays held by each processor. In order to obtain the coarsest partitioning grain possible, the dimension(s) having the largest number of distinct values is chosen to be the partitioning dimension. Assume a dataset with five attributes (A, B, C, D, M), among them A, B, C, D are the dimensions (axis) in the multi-dimensional array, and M is the measure stored in each cell of multi-dimensional array. D_a, D_b, D_c and D_d are the number of distinct values in each dimension, respectively, with $D_a \geq D_b \geq D_c \geq D_d$ established. This dataset will be partitioned and distributed over p processors, numbered $P_0...P_{n-1}$. Thus, a one-dimension partitioning will partition on A, since the A has the biggest number of distinct values[7]; this partitioning also builds an order on A, which means if $A_x \in P_i$ and $A_y \in P_j$ then $A_x \leq A_y$ for $i < j$.

The sub-cubes are constructed over processors with their local sub-dataset (i.e. partition). In order to guarantee that each partition does not have overlaps over the partitioning dimension's distinct values, the sampling-like record distributing methods, such as hash-based or sort-based method can be used to distribute records to various processors as described in [Sanjay and Alok, 1997].

Constructing the sub-cube is performed by scanning the sub-dataset attributed to the local processor. In reference [Sanjay and Alok, 1997], the sub-dataset is scanned twice. The first scan obtains the distinct values for each dimension contained in the sub-dataset, and constructs a hash-table for various dimensions' distinct values. The second scan loads the records into the multi-dimensional array. Record loading (the second scan) works together with probing the hash-tables created earlier. During this process, the method chosen for partitioning and distributing the original dataset will affect the performance because the way to access data is slightly different.

---

[7]Similarly, a two-dimension partitioning will partition on A, B, since A and B have the largest number of distinct values.

Another thing to be noted is that data partitioning determines the amount of data movement during the aggregates' computations of the aggregates [Goil and Choudhary, 1999]. As the computations of various cuboids involve multiple aggregations over any combination of dimensions, some cuboid computations are non-local. They need to partition over a dimension and distribute the partitions once again. Assume that the multi-dimensional array of the 4-dimensional cube is partitioned over A, B, then the aggregation of over dimension C from $\underline{ABC}$[8] to $\underline{AC}$ involves aggregations over dimension B, and requires partitioning and distribution over dimension C.

4.4.2.2   Partitioning star-schema data

In ROLAP, data is organized under the star-schema. Horizontal partitioning was considered an effective method compared to vertical partitioning for star schema data [Ladjel et al., 1999]. In the centralized Data Warehouse, data is stored in a form of a star schema. In general, a star schema is composed of multiple dimension tables and one fact table. Since horizontal partitioning addresses the issue of reducing irrelevant data access, it is helpful to avoid unnecessary I/O operations. One of the features of data analysis queries run on Data Warehouse is that they involve multiple join operations between dimension tables and the fact table. The derived horizontal partitioning, developed for optimizing relational database operations, can be used to efficiently process these join operations.

Partitioning only fact table.   This partitioning scheme only partitions the fact table, and replicates the dimension tables, since the fact table is generally large.

The reference [Bernardino and Madeira, 2012] proposed a stripping-partitioning approach. In this approach, the dimension tables are fully replicated over all compute nodes without being partitioned, as they are relatively small. The fact table is partitioned using round-robin partitioning and each partition is distributed to a compute node. Defining N as the number of computers, each computer stores $1/N$ fraction of the total amount of records. Records of the fact table are striping-partitioned by N computers, then queries can be executed in parallel. In this way, they guarantee a near linear speed-up and a significant improvement in query response time.

The size of each partition determines the workload attributed to a processor. The partition size needs to be tuned according to variant queries. A virtual partitioning method [Akal et al., 2002] was proposed for this purpose. It allows greater flexibility on node allocation for query processing than physical data partitioning. In this work, the distributed Data Warehouse is composed of several database systems running independently. Data tables are replicated over all nodes, and each query is broken into sub-queries by appending range predicates specifying an interval on the partitioning key. Each database system receives a sub-query and is forced to process a different subset of data of the same size. However, the boundaries limiting each subset are very hard to compute, and dispatching the one sub-query per node makes

---

[8]The underlined letters represent the dimensions being partitioned and distributed.

it difficult to realize dynamic load balancing. A fine-grained virtual partitioning (FGVP) [Lima et al., 2004b] was proposed to address this issue. FGVP decomposes the original query into a large number of sub-queries instead of the one query per database system. It avoids fully scanning tables and suffers less from the individual database system internal implementation. However, determining appropriate partition size remains difficult. Adaptive Virtual Partitioning (AVP) [Lima et al., 2004a] adopted an experimental approach to obtain the appropriate partitioning size. An individual database system processes the first received sub-query with a given small partitioning size. Each time it starts to process a new sub-query, it increases the partitioning size. This procedure repeats until the execution time does not shorten any more, then the best partitioning size is found. Performing AVP needs some metadata information. Metadata information includes the clustered index of the relations, the names and cardinalities of relations, the attributes on which a clustered index is built, and the range of values of such attributes. The metadata information is stored in a catalog in the work of [Kotowski et al., 2007].

Partitioning dimension tables and the fact table This partitioning scheme works with star-schema and partitions both dimension tables and the fact table. Often, the dimension tables are horizontally partitioned into various fragments, and the fact table is also horizontally partitioned according to the partitioning results of dimension tables. This scheme takes into account the star-join requirements.

The number of the fact table partitions depends on the partition number of each dimension table. Assume N is the number of fact table partitions, $p_1...p_d$ are the partition numbers of dimension tables $1...d$. If fact table partitioning considers all partitioning performed on the dimension table, then $N = p_1 \times ... \times p_d$. That means, along with the augment of $p_1...p_d$, N will increase explosively. The work of [Bellatreche and Boukhalfa, 2005] focuses on finding the optimal number of fact table partitions, in order to satisfy two objectives:

- avoid an explosion in the number of the fact table partitions;

- ensure a good performance of OLAP queries.

A generic algorithm is adopted for selecting a horizontal schema in their work.

4.4.3 Vertical partitioning of a multi-dimensional dataset

A multi-dimensional dataset usually contains many attributes. With the entire record being stored on the disk (in case of horizontal partitioning), the data access over a multi-dimensional dataset may become inefficient, even though some indexing techniques, such as B-tree are applied to it. Vertical partitioning is needed in some special cases. Imagine the following extreme scenario, where a query scans only the values of one particular attribute of each record. Clearly, in this case, scanning the required attribute separately is much more efficient than scanning the whole table. From the literature, we summarized two types of datasets, for which vertical partitioning is very suitable, the high dimension dataset and the read-oriented dataset.

The advantages of vertical partitioning versus horizontal are, first, it can reduce the dimensionality, which in turn enhances the data accessibility. Second, it enables a set of optimization, like index and easy compressing, to be performed, which in turn improves the efficiency.

4.4.3.1   Reducing dimensionality by vertical partitioning

In a dataset with high dimensionality, the number of dimensions is very large, but the number of records is moderate. The queries run over such a dataset only concern several dimensions. Although OLAP queries involve high-dimension space, retrieving data from all dimensions occurs very rarely. Based on this, the authors of reference [Li et al., 2004] employed a vertically partitioning method in their work. They vertically partitioned the dataset into a set of disjoint low dimensional datasets, called fragments. For each fragment, the local data cube is calculated. These local data cubes are on-line assembled when queries concerning multiple fragments need to be processed. In this work, an inverted index-based indexing technique and data compressing technique are applied to accelerate the on-line data cube assembly.

4.4.3.2   Facilitating index and compression by vertical partitioning

In a vertically partitioned dataset, data is stored in a column-oriented style. Unlike in row-oriented storage, where records are stored one after another, in column-oriented storage, attribute values belonging to the same column are stored contiguously, compressed, and densely packed [Abadi et al., 2009]. OLAP applications are generally read-intensive, where the most common operation is to read data from a disk; the update operation also occurs, but not frequently. For such read-intensive applications, the most important performance-affecting factor is the I/O efficiency.

For a multi-dimensional dataset, the traditional indexing techniques, such as B-tree indexing, are not appropriate. Simply scanning the vertically partitioned data tables is often more efficient than using B-tree based indexes to answer ad hoc range queries [Stockinger et al., 2002]. Using the traditional indexing techniques to process queries, involving only a subset of attributes, suffers from the high dimensionality of the dataset, since the size of the index increases super-linearly with the augmenting of the dimension number. With the vertical partitioning method, the high dimensionality issue is resolved. However, in view of the new data storage structure, not all of the traditional indexing techniques are appropriate. Bitmap index data structure is mostly used for answering read-intensive OLAP queries [Chaudhuri and Dayal, 1997], but not optimized for insert, delete, or update operations. Given this characteristic, Bitmap is considered to be the most suitable index for working with vertical partitioned data. For large datasets, a Bitmap index can have millions to billions of bits. It is imperative to compress the bitmap index. The authors Stockinger et al. [Stockinger et al., 2002] have compared some of the compression schemes such as Byte-aligned Bitmap Code (BBC), Word-Aligned Hybrid run-length code (WAH), and Word-aligned Bitmap Code (WBC). They found that WAH is the most efficient in answering queries because it is much more CPU-efficient.

Compared to row-oriented storage, column-oriented storage presents a number of opportunities to improve performance by compression techniques. In such column-oriented storage, compression schemes encoding multiple values at one time are natural. For example, many popular modern compression schemes, such as Run-length encoding, make use of the similarity of adjacent data to compress. However, in a row-oriented storage system, such schemes do not work well, because an attribute is stored as a part of an entire record. Compression techniques reduce the size of data, thus it improves the I/O performance in the following ways [Abadi et al., 2006]:

- In a compressed format, data is stored more closely, thus the seek time is reduced;

- The transfer time is also reduced because there is less data to be transferred;

- The buffer hit rate is increased because a larger fraction of retrieved data fits in the buffer pool.

Especially, compression ratios are usually higher in column-oriented storage because consecutive values of a same column are often quite similar to each other.

4.5   Data replication

The data replication technique is usually used together with data partitioning. Data replication is used to increase the liability. Multiple identical copies of data are stored over different machines. If the machine holding the primary copy is down, then the data can still be accessed on machines holding the copies. In general, data replication and distribution are not necessarily used with data partitioning. These technologies can be used alone. In the work of [Lima et al., 2004a], the author proposed an adaptive virtual partitioning for OLAP query processing based on shared-nothing architecture. In their approach, the dataset is replicated over all the nodes in the shared-nothing cluster. The virtual partitioning does not physically partition the dataset, instead, it creates a set of sub-queries including different predicates. By applying these predicates, the original dataset is virtually partitioned, the original query is run only on the data items belonging to the partition.

4.6   Query processing parallelism

Parallelizing query processing over partitioned datasets using multiple processors can significantly reduce the response time. The query processing parallelism has shown a good speed-up and scale-up for OLTP query and it is worthwhile investigating parallelism research for processing OLAP query. In the sequential database systems, such as the relational database system, queries are often parsed into graphs during the processing. These graphs are called query execution plans or query plans, which are composed of various operators.

A lot of parallel query processing work has been done on parallel database machines, such as Gamma [DeWitt et al., 1986], Bubba [Boral et al., 1990], Volcano [Graefe, 1990] etc. The main contributions of their work were parallelization of data manipulation and the design of the specific hardware. Even though parallel database machines were not really put into use, they led database technology in the right direction, and its research work became the basis of parallel query processing techniques. The general description of query processing parallelization is as follows: a query is transformed into N partial queries that are executed in an independent way on each of N computers. Generally, we can distribute the same query to all computers, but some types of queries require rewriting.

4.6.1   Inter- and intra-operators

There are several forms of parallelism that are of interest to designers and implementers of query processing systems [Graefe, 1993]. *Inter-query parallelism* means multiple queries are processed concurrently. For example, several queries contained in a transaction are executed concurrently in a database management system. For this form of parallelism, resource contention is an issue. Based on the algebraic operators parallelization, the parallelism forms can be further refined.

Inter-operator parallelism.   Inter-operator parallelism means parallel execution of different operators in a single query. It has two sub-forms, i.e. *horizontal inter parallelism* and *vertical inter parallelism*. *Horizontal inter parallelism* means splitting a tree of query execution plan into several sub-trees, each sub-tree is executed by a processor individually. It can easily be implemented by inserting a special type of operator, *exchange*, into the query execution plan, in order to parallelize the query processing. We will talk about the *exchange* operator in the following content. *Vertical inter-parallelism* is also called *pipeline*, in which operators are organized into a series of *producers* and *customers*. Parallelism is gained by processing records as a stream. Records being processed by producers are sent to customers. The authors of reference [DeWitt and Gray, 1992] argued that, in a relational database system, the benefit of pipeline parallelism is limited. The main reasons were: 1) very long pipelines are rare in query processing based on SQL operators; 2) some SQL operators do not emit the first item of output until they have consumed all items of input, such as aggregate and sort operators; 3) there is often one operator which takes much longer than other operators, which makes the speed-up by pipeline parallelism very limited.

Intra-operator parallelism.   Another form of parallelism is *Intra-operator parallelism*, which means executing an operator using several concurrent processes running on different processors. It is based on data partitioning. The precondition of *intra operator parallelism* is that queries should focus on *sets*. Otherwise, if data being queried represents a sequence, for example, time sequence in a scientific database, then such a form of parallelism could not be directly used, and some additional synchronization should be processed at the result-merging phase.

4.6.2   Exchange operator

The *exchange* operator was proposed in the Volcano system [Graefe, 1990]. It is a parallel operator inserted into a sequential query execution plan so as to parallelize the query processing. It is similar to the operators in the system, like *open*, *next*, *close*; other operators are not affected by the presence of *exchange* in the query execution plan. It does not manipulate data. On the logical level, *exchange* is "no-op," that is, has no place in logical query algebra such as the relational algebra. On the physical level, it provides the "control" functions that the other operators do not offer, such as, processes management, data redistribution, and flow control. *Exchange* only provides "control" parallelisms, but does not determine or presuppose the policies applied for using these mechanisms, such as degree of parallelism, partitioning functions, or attributing processes to processors. In Volcano, the optimizer or user determines these policies. Figure 4.1 shows a parallel query execution plan with *exchange* operators.

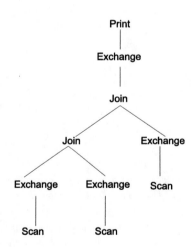

FIGURE 4.1: Parallel query execution plan with *exchange* operators.

4.6.3   SQL operator parallelization

Query running over the partitioned dataset can achieve parallelism, which is also called partitioned parallelism. The algorithms used to implement various operators in parallel are different from those used in a sequential query execution plan implementations. In the following content, we will summarize the parallelization issues for different operators.

Various SQL operators parallelization algorithms have been introduced in the literature [DeWitt and Gray, 1992, Graefe, 1993, Kossmann, 2000], such as, *parallel scan, parallel selection* and *update, parallel sorting, parallel aggregation and duplicate removal, parallel join and binary matching*. Apart from these traditionally used SQL operators, some operators specifically designed for parallel query processing, such as *merge* and *split*, are also introduced. We summarize these algorithms in this section.

Parallel scan.   *Scan* is a basic operator used in query processing. It involves a large number of disk I/Os, which is also the most expensive operation. Therefore, it is significant to parallelize *scan* operator in order to share I/O cost. After partitioning data, each *parallel scan* operator performs over one partition. The output of *parallel scans* working over partitions of a same relation are then processed by a *merge* operator, which merges multiple scanning outputs into one output and sends it to the application or to the next operator in the query execution plan.

Merge and split.   A *merge* operator is used to collect data. A *merge* operator is equipped with several input ports and one output port. The input data streams are received at the input ports of a *merge* operator, and the merging result exits from the output port. If multi-stage parallel processing is required, then a data stream needs to be split into individual sub-streams.

A *split* operator serves this purpose. *Split* is used to partition or duplicate a record stream into multiple ones. For example, a record's various attributes are sent to different destination processes through the attributed *split* operator. A *split* operator partitions the input record stream by applying round-robin, hash partitioning methods, or any other partitioning methods. The *split* allows the auto parallelism of a newly added system operator, and it supports various kinds of parallelism.

Parallel selection and update.   *Parallel selection* operators partition the workload of selection over several I/O devices, each being composed of one single disk or an array of disks. *Selection* operators concurrently perform over all required data partitions, and retrieve matching records. If the partitioning attribute is also the selection attribute, then all disks holding partitions will not contain the selection results. Thus, the number of processes and that of activated disks are limited. Local indexing can still result in high efficiency for a *parallel selection* operator.

Data movement could be caused by an *update* operator updating the value of the partitioning attribute of one record. The modified data might need to be moved to a new disk or node in order to maintain the partitioning consistency. Since moving data is an expensive operation, it is more practical to choose an immutable attribute such as the partitioning attribute in cases where the original dataset contains dynamic data.

Parallel sorting. *Sorting* is one of the most expensive operators in database systems. A lot of research has addressed *parallel sorting*. Without loss of generality, assume a *parallel sorting* operator with multiple inputs and multiple outputs, and further, assume records are aligned in a random order on the sorting attribute in each input, and the output has to be range-partitioned with records being sorted within each range. The algorithms implementing *parallel sorting* generally include two phases, the local sorting phase and the data exchange phase. In the local sorting phase, records are sorted within multiple processes. In the data exchange phase, records are sent to a set of processes. The target process, to which a record is sent, will produce an output partition with the range of sorting attribute value comprising the record's sorting attribute value. In other words, the sent records should contribute to the output produced by the target process. In practice, we can first run data exchange, then local sorting, or, run local sorting first, then data exchange. If data exchange runs first, then the knowledge of quantile should be available in order to ensure load balancing. If local sorting runs first, records are sent, at the end of local sorting, to the right receiving processes, according to the range that each sent record's sorting attribute value belongs to.

One of the possible problems during this procedure is deadlock. The reference [Graefe, 1993] summarized the five necessary conditions of deadlock, cited as follows, i.e. if all these conditions establish, then deadlock will occur. Assuming that a couple of *parallel sort* operators play with other operators in a relationship of producers and consumers, then the necessary conditions of deadlock are:

- multiple consumers feed multiple producers;

- each producer produces a sorted stream and each consumer merges multiple sorted streams;

- some key-based partitioning rule, i.e., hash partitioning, is used other than range partitioning;

- flow control is enabled;

- the data distribution is particularly unfortunate.

Deadlock can be avoided by guaranteeing any one of the above conditions does not establish. Among these, the second condition—each producer produces a sorted stream and each consumer merges multiple sorted streams—is most easily avoided. For instance, if the sending process (producer) does not perform sorting, or each individual input stream of receiving process (consumer) is not sorted, then deadlock can be avoided. That is, moving the sorting operation from producer operator to consumer operator can resolve the deadlock problem.

Deadlocks can also occur during the execution of a *sort-merge-join*, and they can be similarly avoided by moving the sorting operation from the producer operator to the consumer operator. However, this happens when it is not possible to move the sorting operation from producer (sort) to consumer (merge-join), for example, reading data from a B-tree index sorts the records when they are retrieved from the disk. In such a case, it is necessary to find alternative methods that do not require re-partitioning and merging of sorted data between the producers and consumers. The first alternative method is moving the consumers' operations into the producers. Assume the original data is sorted and partitioned with the range- or hash-partitioning method, with the partitioning attribute being exactly the same as the attribute considered by the operations of consumer process, e.g., join attribute in case of consumer operator being merge-join, then the process boundaries and data exchange can be entirely removed from the consumer. This means producer operator, B-tree scan and consumer operator, merge-join, are all performed in a same group of processes. The second method utilizes fragment-and-replicate to perform a join operation. Assume that records of input stream are sorted over a relevant attribute within each partition, but partitioned either round-robin or over a different attribute. For such a data distribution, fragment-and-replicate strategy is applicable. During the join operation with fragment-and-replicate strategy, one input of join is partitioned over multiple processes and another input of join is replicated across these processes[9]. The join operations are running within the same processes as those producing sorted output. Thus, the sorting and join operations are running in one operator, and deadlocks can be avoided.

Parallel aggregation and duplicate removal.   There are three commonly used methods for parallelizing *aggregation and duplicate removal*. The *Centralized Two Phase* method first does aggregations on each of the multiprocessors over the local partition, then the partial results are sent to a centralized coordinator node, which merges these partial results and generates the final result. The *Two Phase* method parallelizes the processing of the second phase of the *Centralized Two Phase* method. The third method is called *Re-partitioning*. It first redistributes the relation on the group by attributes, and then does the aggregation and generates the final results in parallel over each node. Shatdal et al. [Shatdal and Naughton, 1995] argued that these three methods do not work well for all queries. Both *Two Phase* methods only work well when the number of result records is small, whereas the *Re-partitioning* method works well only when the number of distinct values of group-by attributes is large. They proposed a hybrid method that changes/decides the method according to the workload and the number of the distinct values of group-by attributes being computed. A bucket overflow optimization of *Two Phase* methods was discussed in [Graefe, 1993]. For hash-based aggregation, a special technique which improves

---

[9]In typical fragment-and-replicate join processing, the larger input is partitioned, and the smaller input is replicated.

performance is that they do not create the overflow file[10], and the records can be moved directly to the final nodes, because resending records to other nodes is faster than writing records on to a disk. The disk I/O operations are caused when the aggregate output is too large to fit into memory.

Parallel join. *Join* operators include different kinds of join operations, which are performed in different approaches. For instance, *semi-join, outer-join, non-equi-join* etc. are all *join* operators. Unlike the operators mentioned previously, join operators are binary operators, which involve two inputs.

Executing distributed join operators in parallel indispensably involves *send* and *receive* operations. These operations are based on protocols like TCP/IP or UDP. *Row blocking* is a commonly used technique for shipping records to reduce cost. Record shipping is done in a block-wise way, i.e. instead of being shipped one by one, records are shipped block by block. This method compensates the arrival of data up to a certain point [Kossmann, 2000].

Parallel joins over horizontally partitioned data can be achieved in multiple ways. Assuming relation R is partitioned into R_1 and R_2: $R = R_1 \bigcup R_2$, then the join between relations R and S can be computed by $(R_1 \bigcup R_2) \bowtie S$ or $(R_1 \bigcup S) \bowtie (R_2 \bigcup S)$. If R is partitioned into three partitions, and S is replicated, then more methods can be adopted. For instance, the join can be calculated by $((R_1 \bigcup R_2) \bowtie S) \bigcup (R_3 \bowtie S))$, with one replica of S placed near R_1 and R_2, and another replica of B placed near R_3. If $R_i \bowtie S_j$ is estimated to be, then this partial calculation can be removed to reduce the overhead.

Sort-merge-join is a conventional method for computing joins. Assume R and S are still two input relations for join. In *sort-merge-join* method, both of the input relations are first sorted over the join attribute. Then these two sorted intermediate relations are compared, and the matching records are output. *Hash-join* is an alternative to *sort-merge-join* and *Hash-join* breaks a join into several smaller joins. The two input relations R and S are hash-partitioned on the join attributes. One partition of relation R is hash into memory, the related partition of relation S is scanned. Each of the records in this S partition is compared with the partition of R held in memory. Once a record is matched, it is output. *Double-pipelined-hash-join* improved the conventional *hash-join*. It is a symmetric, incremental join. *Double-pipelined-hash-join* creates two in-memory hash-tables, for each one of the input relations. Initially, both hash-tables are empty. The records of R and S are processed one by one. To process one record of R, the hash-table of S is probed; if the record is matched records, then it is outputed immediately. Simultaneously, the record is inserted into the hash-table of R for matching the unprocessed records of S. Thus, at any point in time, all the encountered records are joined. *Double-pipelined-hash-join* has two advantages; first, it allows delivery of the first results of a query as early as possible. Second, it makes it possible to fully exploit pipelined parallelism, and in turn reduce the overall execution time.

---

[10]Overflow file means the common overflowing zone of hash table.

Symmetric partitioning and *fragment-and-replicate* are two basic techniques for parallelizing binary operators. In *symmetric partitioning*, both inputs are partitioned over the join attribute, and then the operations will be run on every node. This method is used in Gamma [DeWitt et al., 1986]. In the *fragment-and-replicate* method, one of the two inputs is partitioned; the other input is broadcast to all other nodes. In general, the larger input is partitioned in order not to move it. This method was realized in the early database systems, because the communication cost overshadowed the computation cost. Sending small input to a small number of nodes costs less than partitioning both larger input and small input. To be noted, *fragment-and-replicate* cannot work correctly for *semi-join*, and other binary operators, like, *difference union*, because when a record is replicated, it will contribute multiple times to the output.

Semi-join is used to process join between relations placed on different nodes. Assume two relations R and S are placed on nodes r and s, respectively. *Semi-join* sends the needed columns for join of relation R from node r to s, then finds the records qualifying the join from relation S and sends these records back to r. The join operation is executed on node r. *Semi-join* can be expressed as: $R \bowtie S = R \bowtie (S \ltimes_\pi (R))$. *Redundant-semi-join* is a technique for reducing network traffic used in distributed databases for join processing. This method is used in distributed memory parallel systems. Assume two relations R and S having a common attribute A, are each stored on nodes r and s separately. *Redundant-semi-join* sends the duplicate-free projection on A to s, executes a *semi-join* to decide which records of S will contribute to the join result, and then ships these records to r. Based on the law of relational algebra $R \bowtie S = R \bowtie (S \ltimes R)$, there is no need for shipping S, which reduces communication overhead, at the cost of adding the overhead of projecting, shipping the column A of R and executing the semi-join. Such a reduction can be applied on R or S, or both. The operations included during this process, such as *projection*, *duplicate removal*, *semi-join* and *final join* can be parallelized not only on nodes s and r, but also on more than two nodes.

Symmetric fragment-and-replicate is proposed by Stamos et al. [Stamos and Young, 1993] which is applicable for *non-equi-joins* and *N-way-join*. For parallelizing a *non-equi-join*, processors are organized into rows and columns. One input relation is partitioned over rows, and its partitions are replicated over each processor row. The other input relation is partitioned over columns, and its partitions are replicated over each processor column. A record of one input relation only matches with one record from the other input relation. The global join result is the concatenation of all partial results. This method improves the *fragment-and-replicate* method by reducing the communication cost.

For joins in a parallel Data Warehouse environment, the *parallel star-join* is discussed by Datta et al. in [Datta et al., 1998]. This parallel join processing is based on a particular data structure, *Data Index*[11], proposed in the same work. Recall that *Basic DataIndex* (*BDI*) is simply a vertical partition of the fact table, which may in-

---

[11]DataIndex is discussed in Section 4.3

clude more than one column and the *Join DataIndex* (*JDI*) is designed to efficiently process join operations between the fact table and the dimension table. *JDI* is an extension of *BDI*. *JDI* is composed of *BDI* and a list of RecordIDs indicating the matching records in the corresponding dimension table. Assume that F represents the fact table, and D represents the set of dimension tables, then $\mid D \mid = d$, which means that there are d dimension tables. Let G represent a set of processor groups, and $\mid G \mid = d + 1$. Dimension table D_i and the fact table partition *JDI* corresponding to the key value of D_i are distributed to the processor group i. And the fact table partition *BDI* after containing measures is distributed to processor group $d + 1$. Based on the above data distribution, the *parallel star-join* processing only involves rowsets and projection columns.

In [Akinde et al., 2003] a more complex join operator, *General Multiple Dimension Join* (GMDJ), is discussed in a distributed Data Warehouse environment. GMDJ is a complex OLAP operator composed of relational algebraic operators and other GMDJ operators. These GMDJ operator-composed queries need multi-round processing. GMDJ operator clearly separates group-by definition and aggregate definition, which allows to express various kinds of OLAP queries. An OLAP query expressed in GMDJ expressions is translated into a multi-rounded query plan. During each round, each site of distributed Data Warehouse executes calculations and communicates its results to the coordinator; the coordinator synchronizes the partial results into a global result, then transfers the global result to distributed Data Warehouse sites. When a distributed Data Warehouse site receives an OLAP query, it transforms the OLAP query into GMDJ operators, which are then optimized using distributed computation. Taking the efficiency into account, the synchronization at the end of each round is started when the faster sites' partial results arrive on the coordinator, instead of waiting for all partial results arrivals before starting the synchronization. Although this work described how to generate a distributed query plan, it did not support on-line aggregation.

4.6.3.1 Issues of query parallelism

During the parallelization of query processing, some issues will appear, such as *data skew* and *load balance*. Pipeline parallelism does not easily lend itself to load balancing, since each processor in the pipeline is loaded proportionally to the amount of data it has to process. This amount of data cannot be predicated very well. For partitioning-based parallelism, load balancing is optimal, if the partitions are all of equal size. However, load balancing can be hard to achieve in the case of data skew. Range partitioning risks *data skew*, where all the data is placed in one partition, and all the calculations. However, hashing and round-robin based partitioning suffer less from data skew.

4.7   Concluding remarks

Recently, some new technologies have been adopted in the multi-dimensional data analysis application. These new technologies include: in-memory query processing, search engine technologies, and enhanced hardware. Commercial multi-dimensional data analysis products use these new technologies and commonly run on a couple of blade servers. They do not adopt the traditional pre-calculation method, i.e. storing pre-calculated results into materialized views, to accelerate query processing. On the contrary, they compress data to fit into the memory. Before a query is answered, all the data needed to answer it is copied to memory. All the query processing is in-memory.

Aside from in-memory query processing, commercial multi-dimensional data analysis products adopted the search engine technology to accelerate the query processing. They used a **metamodel** in order to bridge the gap between the structured data cube and search engine technology, which was originally developed to work with unstructured data. In this metamodel, the data originally stored under star-schema is represented as a join graph expressing the joins between fact table and required dimension tables.

In this chapter, we first described the features of multi-dimensional data analysis queries and three distributed system architectures, including shared-memory, shared-disk, and shared-nothing. Second, we gave a survey of existing work on accelerating data analytical query processing. Three approaches were discussed, pre-computing, data indexing, and data partitioning. Pre-computing is an approach to bartering storage space for computing time. The aggregates of all possible dimension combinations are calculated and stored to rapidly answer the forthcoming queries. We discussed some related issues of pre-computing, including data cube construction, sparse cube, query result re-usability, and data compressing. For indexing technologies, we discussed several indexes appearing in the literature, including B-tree/B^+-tree index, projection index, Bitmap index, Bit-Sliced index, join index, inverted index etc. A special type of index used in distributed architecture was also presented. For data partitioning technology, we introduced two basic data partitioning methods, horizontal partitioning and vertical partitioning, as well as their advantages and disadvantages. Then we presented the application of partitioning methods on the multi-dimensional dataset. After that, the parallelism of query processing was described. We focused on parallelization of various operators, including *scan*, *merge*, *split*, *selection*, *update*, *sorting*, *aggregation*, *duplicate removal*, and *join*. At the end of this chapter, we introduced some new developments in multi-dimensional data analysis.

Chapter 5

Data intensive applications with MapReduce

5.1 Introduction

Along with the development of hardware and software, more and more data is generated at a rate much faster than ever. Although data storage is inexpensive, and the issues of storing large volumes of data can be resolved, processing large volumes of data is becoming a challenge for data analysis software. The feasible approach handling large-scale data processing is to *divide and conquer*. People look for solutions based on a parallel model, for instance, the parallel database, which is based on the shared-nothing distributed architectures. Relations are partitioned into pieces of data, and the computations of one relational algebra operator to be proceeded in parallel on each piece of data [DeWitt and Gray, 1992]. The traditional parallel attempts in data intensive processing, like parallel database, were suitable when data scale was moderate. However, parallel database does not scale well. MapReduce is a new parallel programming model, which turns a new page in data parallelism history. The MapReduce model is a parallel data flow system that works through data partitioning across machines, each machine independently running the single-node logic [Hellerstein, 2012]. MapReduce initially aims at supporting information preprocessing over a large number of web pages. MapReduce can handle large datasets with the guarantee of scalability, load balancing, and fault tolerance, and MapReduce is applicable to a wide range of problems. Depending on different problems, the detailed implementations are varied and complex. The following are some of the possible problems to be addressed:

- How to decompose a problem into multiple sub-problems

- How to ensure that each sub-task obtain the data it needs

- How to cope with intermediate output so as to benefit from MapReduce's advantages without losing efficiency

- How to merge the sub-results into a final results

In this chapter, we will focus on the MapReduce model. First, we describe the logical composition of the MapReduce model as well as its extended model. The issues related to this model, such as MapReduce's implementation frameworks, cost analysis, etc., will also be addressed. Second, we will talk about the distributed data access of MapReduce. A general presentation on data management applications in the cloud is given before the discussion about large-scale data analysis based on MapReduce.

5.2   MapReduce: New parallel computing model in cloud computing

In parallel distributed computing, the most troublesome part of programming is handling the system-level issues, such as communication, error handling, synchronization etc. Some parallel computing models, such as MPI, OpenMP, RPC, RMI etc., are proposed to facilitate the parallel programming. These models provide a high-level of abstraction and hide the system-level issues, like communication and synchronization issues.

Message Passing Interface (MPI) defines a two-sided message-passing library (between sender and receiver). Otherwise, one-sided communication is also possible. Note that in MPI, a send operation does not necessarily have an explicit reception. Remote Procedure Call (RPC) and Remote Method Invocation (RMI) are based on the one-side communication. Open Multi-Processing (OpenMP) is designed for shared memory parallelism. It automatically parallelizes programs, by adding the synchronization and communication controls during compiling time. Although these models and their implementations have undertaken much system-level work, they are rather designed for realizing processor-intensive applications. When using these models for large-scale data processing, programmers still need to handle low-level details.

MapReduce is a data-driven parallel computing model proposed by Google. The first paper on the MapReduce model [Dean and Ghemawat, 2004] described one possible implementation of this model based on large clusters of commodity machines with local storage. The paper [Lämmel, 2007] gave a rigorous description of this model, including its advantages, in Google's domain-specific language, Sawzall. One of the most significant advantages is that it provides an abstraction which hides the system-level details from programmers. Having this high-level abstraction, developers do not need to be distracted by solving how computations are carried out and finding the input data that the computations need. Instead, they can focus on the processing of the computations.

5.2.1   Dataflow model

MapReduce is a parallel programming model proposed by Google. It aims at supporting distributed computation on large datasets by using a large number of computers with scalability and fault tolerance guarantees. During the map phase, the master node takes the input, and divides it into sub-problems, then distributes them to the worker nodes. Each worker node solves a sub-problem and sends the intermediate results ready to be processed by reducer. During the reduce phase, intermediate results are processed by reduce function on different worker nodes, and the final results are generated.

This type of computation is different from parallel computing with shared memory, which emphasizes that computations occur concurrently. In parallel computing with shared memory, the parallel tasks have close relationships with each other. Computations supported by MapReduce are suitable for parallel computing with distributed memory. Indeed, MapReduce executes the tasks on a large number of distributed computers or nodes. However, there is a difference between the computations supported by MapReduce and the traditional parallel computing with distributed memory. For the latter, the tasks are independent, which means that the error or loss of results from one task does not affect the other tasks' results, whereas in MapReduce, tasks are only relatively independent and loss or error do matter. For instance, the mapper tasks are completely independent of each other, but the reducer tasks cannot start until all mapper tasks are finished, i.e. reducer tasks' start-up is restricted. The loss of task results or failed execution of tasks also produce a wrong final result. With MapReduce, complex issues such as fault-tolerance, data distribution and load balancing are all hidden from the users. MapReduce can handle them automatically. In this way, MapReduce programming model simplifies parallel programming. This simplicity is retained in all frameworks that implement MapReduce model. By using these frameworks, the users only have to define two functions, *map* and *reduce*, according to their applications.

Fundamentals of the MapReduce model. The idea of MapReduce was inspired by high-order function and functional programming. *Map* and *reduce* are two primitives in functional programming languages, such as Lisp, Haskell, etc. A *map* function processes a fragment of a key-value pairs list to generate a list of intermediate key-value pairs. A *reduce* function merges all intermediate values associated with a same key, and produces a list of key-value pairs as output. Refer to the reference [Dean and Ghemawat, 2004] for a more formal description. The syntax of the MapReduce model is the following:

map(key1,value1) $\rightarrow$ list(key2,value2)

reduce(key2,list(value2) $\rightarrow$ list(key2,value3)

In the above expressions, the input data of the *map* function is a large set of (key1,value1) pairs. Each key-value pair is processed by the *map* function without depending on other peer key-value pairs. The map function produces another pair of key-values, noted as (key2,value2), where, the key (denoted as key2) is not the original key as in the input argument (denoted as key1). The output of the map phase is processed before entering the reduce phase, that is, key-value pairs (key2,value2) are grouped into lists of (key2,value2), each group having the same value of key2. These lists of (key2,value2) are taken as input

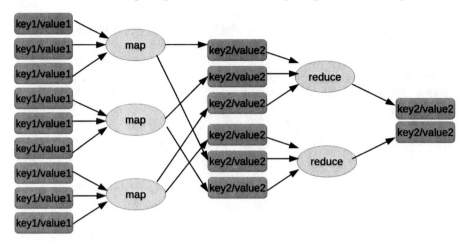

FIGURE 5.1: Logical view of the MapReduce model.

data by the *reduce* function, and the *reduce* function calculates the aggregate value for each key2 value. Figure 5.1 shows the logical view of MapReduce.

The formalization given in the first article of MapReduce [Dean and Ghemawat, 2004] was simplified. It omitted the detailed specification for the intermediate results processing part in order to hide the complexities from the readers. However, this might cause some confusion. The author of reference [Lämmel, 2007] took a closer look at Google's MapReduce programming model and gave a clearer explanation for the underlying concepts of the original MapReduce. The author formalized the MapReduce model with the functional programming language, Haskell. The author also analyzed the parallel opportunities existing in MapReduce model and its distribution strategy. The parallelization may exist in the processing of mapper's input, the grouping of the intermediate output, the reduction processing over groups and the reduction processing inside each group during the reduce phase. In the strategy of the MapReduce model, network bandwidth is considered as the scarce resource. This strategy combines parallelization and large dataset distributed storage to avoid saturating the network bandwidth.

Note that the keys used in the map phase and the reduce phase can be different, i.e. developers are free to decide which part of data will be keys in these two phases. That means this data form of key-value pair is very flexible, which is very different from the intuitive feel. As keys are user-definable, one can ignore the limitation of key-value. Thus, a whole MapReduce procedure can be informally described as follows:

- Read a lot of data;

- Map: extract useful information from each data item;

- Shuffle and sort;

- Reduce: aggregate, summarize, filter, or transform;

- Write the results.

Extended MapCombineReduce model. The MapCombineReduce model is an extension of the MapReduce model. In this model, an optional component, namely the **combiner**, is added to the basic MapReduce model. This combiner component is proposed and adopted in the Hadoop project [Hadoop, 2012a]. The intermediate output key-value pairs are buffered and periodically flushed onto disk. At the end of the processing procedure of the mapper, the intermediate key-value pairs are already available in memory. However, these key-value pairs are not written into a single file. These key-value pairs are split into *R* buckets based on the key of each pair. For the sake of efficiency, we sometimes need to execute a reduce-type operation within each worker node. Whenever a reduce function is both associative and commutative, it is possible to "pre-reduce" each bucket without affecting the final result of the job. Such a "pre-reduce" function is referred to as a **combiner**. The optional combiner component collects the key-value pairs from the memory. Therefore, the key-value pairs produced by the mappers are processed by the combiner instead of being written into the output immediately. In this way, the intermediate output amount is reduced. This makes sense when the bandwidth is relatively small and the volume of data transferred over the network is large. Figure 5.2 shows the logical view of the MapCombineReduce model.

5.2.2   Two frameworks: GridGain versus Hadoop

Hadoop [Hadoop, 2012a] and GridGain [GridGain, 2012] are two different open-source implementations of MapReduce. Hadoop is designed for processing applications. The response time is relatively long, for instance, from several minutes to several hours. One example of such an application is the finite element method calculated over a very large mesh. The application consists of several steps, each step using the data generated by the previous steps. The processing of Hadoop includes transmitting the input data to the computing nodes. This transfer must be extremely fast to fulfill the users' needs. Hadoop is an excellent MapReduce supporting tool and a Hadoop cluster gives high throughput computing. However, it has a high latency since Hadoop is bound with the Hadoop distributed file system (HDFS). The Hadoop's MapReduce component operates on the data or files stored on HDFS, and these operations take a long time to be performed. For this reason, Hadoop cannot provide a low latency.

However, what we are trying to perform in parallel is a great number of queries on one large dataset. The dataset involved is not modified, and the query processing should be interactive. In fact, low latency is essential for interactive applications. In order to be compatible with the application's interactive requirements, the response time is strictly limited, for instance, within five seconds. In contrast to Hadoop, GridGain is not bound with a file system and offers low latency. It is a MapReduce computational tool. GridGain splits the computing task into small jobs and executes

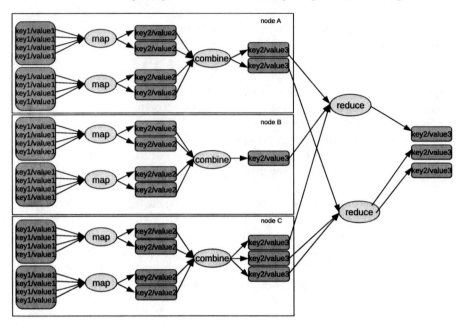

FIGURE 5.2: Logical view of a MapCombineReduce model.

them on the grid in parallel. During the task execution, GridGain deals with the low-level issues, such as nodes discovery, communication, jobs collision resolution, load balancing, etc. When compared with Hadoop, GridGain is more flexible. Instead of accessing data stored on a distributed file system, GridGain can process data stored in any file system or database. In addition, GridGain has some other advantages. For instance, it does not need application deployment and can be easily integrated with other data grid products. In particular, it allows programmers to write their programs in pure Java language.

5.2.3   Communication cost analysis

In parallel programming, a computation is partitioned into several tasks, which are allocated to different computing nodes. The communication cost issues must be considered, since the data transmission between the computing nodes represents a non-negligible part. The communication cost is directly linked with the degree of parallelism. If the tasks are partitioned with a high degree of parallelism, the communication cost will be large. On the other hand, if the degree of parallelism is small, the communication cost will be limited.

In the MapReduce parallel model, the communication cost exists in several phases. For the basic MapReduce model, without a combiner component, the communication cost consists of three distinct phases. The first phase is the launching phase, during which all the tasks are sent to the mappers. The second phase, located between the mappers and reducers, consists in sending the output from mappers to reducers. The third phase is the final phase, which produces the results, and where the outputs of the reducers are sent back. For the extended MapCombineReduce model, the communication consists of four phases. The first phase is still the launching phase. The second phase, located between the mappers and combiners, consists of sending the intermediate results from the mappers to combiners located on the same node. The third phase, located between the combiner and the reducer, consists of sending the output of the combiners to reducers. The fourth phase is the final phase, which produces the results. The size of the output data exchanged between the components strongly impacts the communication cost. In reference [Hasan, 1996], the author described an analysis of the communication cost in a parallel environment, depending on the amount of data exchanged between the processes. Based on their work we analyzed the case of MapReduce, and we summarized the following factors influencing the communication cost.

(i) The first one is the amount of intermediate data to be transferred, from the mappers to the reducers (case without a combiner component) or from the combiners to the reducers (case of a combiner component).

(ii) The second factor is the physical locations of the mappers, the combiners and the reducers. If two communicating components are on the same node, the communication cost is low; otherwise the cost is high. If two communicating components are located on two geographically distant nodes, the communication cost could be extremely high.

(iii) The third factor to be considered is the number of mappers, combiners, and reducers respectively. Usually, the user defines the number of mappers according to the scale of the problem to be solved and the computing capacity of the hardware. The number of combiners is usually equal to the number of nodes participating in the calculation, collecting local intermediate results of a node. Whether or not the number of reducers can be user-definable depends on the design of the implemented MapReduce framework. For example Hadoop allows the user to specify the number of reducers. On the contrary, GridGain fixes the value of the number of reducers to one.

(iv) The fourth factor is the existence of a direct physical connection between two communicating components. A direct physical connection between two components means that two nodes respectively holding the two components are physically connected to each other.

(v) The last factor is the contention over the communicating path. When two or more communications are executed at the same time, the contention of the bandwidth will appear. A possible scenario of this contention with the MapReduce model could be described as follows. The mappers on various nodes are started at almost the same time. Since the nodes in a cluster are usually of identical type, they have almost the same capability. As a consequence, the mappers complete their work on each node at the same time. The outputs of these mappers are then sent to the reducers. In this scenario, the contention of the communicating path is caused by the transmission requests arriving almost simultaneously.

Since the actions of transferring the data from the master node to the worker nodes are generally much more costly than the actions of transferring the mappers from the master to the workers, we usually transfer the mapper job code toward the location of data. Thus, the geographical locations of the data have a strong impact on the efficiency.

5.3   Distributed data storage underlying MapReduce

The data-driven nature of MapReduce requires a specific underlying data storage support. High-Performance Computing's traditional separating storage component from computations is not suitable for processing a large dataset. MapReduce abandons the approach of separating computation and storage. In the runtime, MapReduce needs to either access data on a local disk, or access data stored close to the computing node.

5.3.1   Google file system

Google uses a distributed Google File System (GFS) [Ghemawat et al., 2003], [Passing, 2012] to support MapReduce computations. Hadoop provides an open-source implementation of GFS, which is named the Hadoop Distributed File System (HDFS) [Hadoop, 2012b]. In Google, MapReduce is implemented on top of GFS and is run over within clusters. The basic idea of such a GFS is to divide a large dataset into chunks, then replicate each chunk across different nodes. The chunk size is much larger than in the traditional file system. The default chunk size is 64M in GFS and HDFS.

The architecture of GFS follows the master-slave model. The master node is responsible for maintaining file namespace, managing and monitoring the cluster. Slave nodes manage their actual chunks. Data chunks are replicated across slave nodes, with three replicas by default. When an application wants to read a file, it needs to consult the metadata information about chunks by contacting the master node to know on which slave nodes the required chunk is stored. After that, the application contacts the specific slave nodes to access data. The size of chunk is a crucial factor influencing the amount of data that the master node needs to handle. The default chunk size considers a trade-off between trying to limit resource usage and master interaction times on the one hand, and accepting an increased degree of internal fragmentation on the other hand.

GFS is different from the general applicative File System, such as NFS or AFS, in that it assumes that data is updated in an append-only fashion [Brewer, 2005], and data access is mainly long streaming reads. GFS is optimized for workload characterized by the above-mentioned features. The following summarizes GFS's characteristics:

- GFS is optimized for the usage of large files, where the space efficiency is not very important;

- GFS files are commonly modified by being appended data;

- Modifying at file's arbitrary offset is an infrequent operation;

- GFS is optimized for large streaming reads;

- GFS supports great throughput, but has long latency;

- Client's caching techniques are considered to be ineffective.

A weakness of GFS master-slave is the single master node. The master node plays a crucial role in GFS. It not only manages metadata, but also maintains the file namespace. In order to avoid the master node becoming a bottleneck, the master has been implemented using multi-threading and fine-grained locking. Additionally, in order to alleviate the workload of the master node, the master node is designed to only provide metadata to locate chunks, and it does not participate in the following data accessing. Risk of single point of failure is another weakness of GFS. Once the master node crashes, the whole file system will stop working. For handling a master crash, a *shadow masters* design is adopted. The shadow master holds a copy of the newest operation log. When master node crashes, the shadow master provides read-only access of metadata.

5.3.2   Distributed cache memory

The combination of MapReduce and GFS guarantee high throughput, since GFS is optimized for sequential reads and appends on large files. However, such a combination has a high latency. GFS-based MapReduce heavily uses disk, in order to alleviate the effect brought on by failures. However, this produces a large amount of disk I/O operations. The latency for disk access is much higher than that of memory access. In GFS-based MapReduce, memory was not fully utilized [Zhang et al., 2009]. In the GFS open-source implementation, HDFS, reading data also suffers from high latency. Reading a random chunk in HDFS involves multiple operations. For instance, it requires communicating with the master to obtain the data chunk location. If data chunk is not located on the node where the read operation occurs, then that also requires performing data transfer. Each of these operations leads to higher latency [Lin et al., 2009].

The authors of reference [Zhang et al., 2009] argued that small-scale MapReduce clusters, which have no more than a dozen machines, are very common in most companies and laboratories. Node failures are infrequent in clusters of such size. So it is possible to construct a more efficient MapReduce framework for small-scale heterogeneous clusters using distributed memory. The author of the reference [Lin et al., 2009] also proposed the idea of utilizing distributed memory. Both of these works have chosen the open-source distributed in-memory object caching system, *memcached* to provide Hadoop with an in-memory storage.

In the work of [Lin et al., 2009], the whole dataset, in the form of key-value pair, is loaded into *memcached* from HDFS. Once the whole dataset is in *memcached*, the subsequent MapReduce programs access data with the client API of *memcached*. Each mapper or reducer maintains connections to the *memcached* servers. All requests are made in parallel and are distributed across all *memcached* servers.

In the work of [Zhang et al., 2009], only the output of mappers is loaded into *memcached*. The cached mapper's output is attached to a key. Such a key is made up of a mapper ID and its target reducer ID. Once a reducer starts, it checks whether the outputs for it are in *memcached*. If it is the case, then it retrieves them from the *memcached* server.

5.3.3   Data accessing

If a MapReduce framework is used, without being attached to a distributed file system, then data locating needs to be taken charge of by developers. GridGain is such a pure MapReduce computing framework, and it is not attached to any distributed storage system. Although this forces developers to do the low-level work of data locating, to some extent, it provides some flexibility. Data accessing no longer needs to consult the file namespace; data forms other than files can also be the representation of distributed stored data, e.g. data can be stored in a database on each computing node.

GridGain's MapReduce is composed of one master node and multiple worker nodes. GridGain provides a useful mechanism for users to add user properties, which are visible to master. Master nodes can identify worker nodes from the added properties. The additional properties defined by the user can be used for different purposes, such as the logical name of node, role name of node etc. Because we adopted GridGain as the MapReduce framework, we give an example taken from our work. In this work, we added, to each worker node, a property representing the identifier of data fragment stored in the current worker node.

In the manual approach, a data pre-processing phase is indispensable. During this phase, a dataset is divided into blocks, and distributed/replicated across different worker nodes. Using the method mentioned above, a user-defined property representing a data fragment identifier is added to each worker node. Being visible to the master node, this user-defined property is used for locating data chunks. An illustration is given in Figure 5.3. As this manual approach decouples underlying storage from the computations, it provides the possibility to choose various underlying data storages. Mappers can access a third-part data source. Because a developer can personally control data locating, data transfer between worker nodes is less frequent than in the GFS-based MapReduce system. More importantly, the optimization over

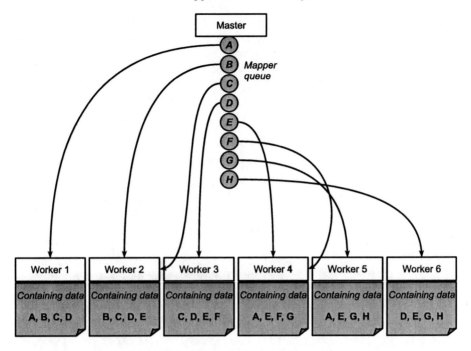

FIGURE 5.3: Manual data locating based on GridGain: circles represent mappers to be scheduled. The capital letter contained in each circle represents the data block to be processed by a specific mapper. Each worker has a user-defined property reflecting the *contained data*, which is visible for the master node. By identifying the values of this property, the master node can locate the data blocks needed by each mapper.

data accessing can be performed without being limited by a particular file system.

5.4 Large-scale data analysis based on MapReduce

Data analysis applications or OLAP applications are encountering scalability issues. Facing more and more generated data, OLAP software should be able to handle much larger datasets than ever. MapReduce naturally has good scalability, and people argued that the MapReduce approach is suitable for a data analysis workload. The key is to choose an appropriate implementation strategy for the given data analysis application. For choosing an appropriate implementation strategy to process a data analysis query, two types of questions need to be answered. The first question is about the data placement. This short term includes several sub-questions: What is the most suitable data-partitioning scheme? To what degree will we partition the dataset? What is the best data placement strategy of data partitions? The second question is how to efficiently perform the query over the distributed data partitions. In a MapReduce based system, the query's calculation is transformed into another problem; how to implement the query's processing with MapReduce. To answer these questions, specific analysis addressing various queries needs to be undertaken.

5.4.1 Data query languages

Hadoop's rudimentary support for MapReduce, promoted the development of MapReduce-based high-level data query languages. A data query language PigLatin [Christopher et al., 2008], was originally designed by Yahoo, and later became an open-source project. It is designed as a bridge between the low-level, procedural style of MapReduce and the high-level declarative style of SQL. It is capable of handling structured and semi-structured data. Programs written in PigLatin are translated into physical plans, composed of MapReduce procedures during compiling. The generated physical plans are then executed over Hadoop.

Similarly, another open-source project, Hive [Hive, 2012] of Facebook, is a Data Warehouse infrastructure built on top of Hadoop. It allows the aggregating of data, the processing of ad hoc queries, and the analysis of data stored in Hadoop files. HiveQL is a SQL-like language, which allows querying over large datasets stored as HDFS files.

Microsoft developed a MapReduce-based declarative and extensible scripting language, SCOPE (Structured Computations Optimized for Parallel Execution) [Ronnie et al., 2008], targeted at massive data analysis. This language is high-level declarative, and the compiler, together with optimizer, can improve SCOPE scripts through compiling and optimizing. SCOPE is extensible. Users are allowed to create customized extractors, processors, aggregators and combiners with the extending built-in C# components.

5.4.2 Data analysis applications

An attempt at a MapReduce-based OLAP system was described in [Chen et al., 2008]. The following description of their work lays out a clear example of doing data analysis with MapReduce. The dataset used in their work is a data cube. In particular, the data cube was coming from a web log, which is employed to analyse the web search activities. More specifically, the data cube is composed of two dimensions (keyword k and time t) and two measures (page count: $pageCount$ and advertisement count: $adCount$). During data partitioning, the data cube is divided over dimensions of keyword and time. The cells having the same value of k and t are put into one block. Similar to the MOLAP, the hierarchy concept is applied over the data cube in their work. These different hierarchy levels in $time$ dimension allow the partitioning of a data cube with different granularities. Support of dynamic data partitioning granularity is a unique feature of this work. Because queries processed in their work are correlated, the results generated for one query can serve as the input for another query. However, the granularity of the second query is not necessarily the same as that of the first query, and then a change of granularity is needed. The MapReduce-based query processing concerns two groups of nodes: nodes running *mappers*, and nodes running *reducers*. The group of nodes is defined at the beginning of query processing. A *mapper* fetches part of dataset and generates *key-value* pairs from an individual record. The *key* field is related to different granularities, which in turn depends on the query, and it can be computed using the given algorithms. The *value* field is the exact copy of the original data record. These *key-value* pairs are shuffled, and dispatched to *reducers*, those with the same *key* going to the same *reducer*. The *reducer* performs an external sorting to group pairs with the same value and then produces an aggregated result for each group.

Regarding commercial software, MapReduce was integrated into some commercial software products. Greenplum is a commercial MapReduce database, which enables programmers to perform data analysis on petabyte-scale datasets inside and outside of it [Greenplum, 2012]. Aster Data Systems, a database software company has recently announced the integration of MapReduce with SQL. Aster's *nCluster* allows one to implement flexible MapReduce functions for parallel data analysis and transformation inside the database [Aster nCluster, 2012].

5.4.3 Comparison with shared-nothing parallel databases

Despite being able to run on different hardware, MapReduce typically runs on a shared-nothing architecture where computing nodes are connected by network, without memory or disk sharing among each other. Many parallel databases adopted shared-nothing architecture, as in the parallel database machines, Gamma [DeWitt et al., 1986] and Grace [Fushimi et al., 1986].

Though MapReduce and parallel databases target different users, it is in fact possible to write almost any parallel processing task as either a set of database queries or a set of MapReduce jobs [Pavlo et al., 2009]. This led to controversies about which system is better for large-scale data processing. Among them, there is also criticism of the new-rising MapReduce. Some researchers in the database field even argued that MapReduce is a step backward in the programming paradigm for large-scale data intensive applications [DeWitt and Stonebraker, 2012]. However, more and more commercial database software has begun to integrate the cloud computing concept into their products. Existing commercial shared-nothing parallel databases suitable for doing data analysis application in clouds are: Teradata, IBM DB2, Greenplum, DATAllego, Vertica and Aster Data. Among others, DB2, Greenplum, Vertica and Aster Data are naturally suitable since their products could theoretically run in the datacenters hosted by cloud computing providers [Abadi, 2009]. It is interesting to compare the features of both systems. We compare a shared-nothing parallel database and MapReduce in the following aspects:

Data partitioning. In spite of having many differences, a shared-nothing parallel database and MapReduce do share one feature: the dataset is all partitioned in both systems. However, as in a shared-nothing parallel database, data is structured in tables, data partitioning being done with specific data partitioning methods. Partitioning takes into account the data semantics, and is run under the control of the user. On the contrary, data partitioning in a typical MapReduce system is automatically done by the system, where a user can only participate in data partitioning with limitations. For example a user can configure the size of the block. But the semantic of the data is not considered during partitioning.

Data distribution. In a shared-nothing parallel database, the knowledge of data distribution is available before query processing. This knowledge can help the query optimizer to achieve load-balancing. In a MapReduce system, the details of data distribution remain unknown, since distribution is automatically done by the system.

Support for schema. Shared-nothing parallel databases require that data conform to a well-defined schema; data is structured in rows and columns. In contrast, MapReduce permits data to be in any arbitrary format. The MapReduce programmer is free of schema, and data can even have no structure at all.

Programming model. Like other DBMSs, a shared-nothing parallel database supports a high-level declarative programming language, i.e. SQL, which is known by and largely accepted by both professional and non-professional users. With SQL, users only need to declare what they want to do, but do not need to provide a specific algorithm to realize it. However, in the MapReduce system, developers must provide an algorithm in order to realize the query processing.

Flexibility. SQL is routinely criticized for its insufficient expressive power. In order to mitigate flexibility, shared-nothing parallel databases allow user-defined functions. MapReduce has good flexibility by allowing developers to realize all calculations in the query processing.

Fault tolerance. Both parallel databases and MapReduce use replication to deal with disk failures. However, parallelism databases cannot handle node failures, since they do not save intermediate results; once a node fails, the whole query processing should be restarted. MapReduce is able to handle node failure during the execution of MapReduce computation. The intermediate results (from mappers) are stored before launching reducers in order to avoid starting the processing from zero in case of node failure.

Indexing. Parallel databases have many indexing techniques, such as hash or B-tree, to accelerate data access. MapReduce does not have built-in indexes.

Support for transactions. The support for transactions requires the processing to respect ACID. Shared-nothing parallel databases support transactions, since they can easily respect ACID. But it is difficult for MapReduce to respect such a principle. Note that, in large-scale data analysis, the ACID is not really necessary.

Scalability. Shared-nothing parallel databases can scale well to tens of nodes, but it is difficult to go any further. MapReduce has very good scalability, which is proved by Google's use. It can scale to thousands of nodes.

Adaptability over a heterogeneous environment. As the shared-nothing parallel database is designed to run in a homogeneous environment, it is not suited to a heterogeneous environment, unlike MapReduce which is able to run in a heterogeneous environment.

Execution strategy. MapReduce has two phases, map phase and reduce phase. Reducers need to pull each input data from the nodes where mappers were run. Shared-nothing parallel databases use a *push* approach to transfer data instead of *pull*. Table 5.1 summarizes the differences between parallel databases and MapReduce with short descriptions.

Table 5.1: Differences between Parallel Databases and MapReduce.

	Parallelism database	**MapReduce**
Data partitioning	Use specific methods consider data semantic	Done automatically do not consider data semantic
Data distribution	Known to developers	Unknown
Schema support	Yes	No
Programming model	Declarative	Direct realize
Flexibility	Not good	Good
Fault tolerance	Handle disk failures	Handle disk and node failures
Indexing	Support	Have no built-in index
Transaction support	Yes	No
Scalability	Not good	Good
Heterogeneous environment	Unsuitable	Suitable
Execution strategy	Push mode	Pull mode

Hybrid solution. MapReduce-like software, and shared-nothing parallel databases have their own advantages and disadvantages. People look for a hybrid solution that combines the fault tolerance, heterogeneous cluster, and ease of scaling of MapReduce and the efficiency, performance, and tool plug-ability of a shared-nothing parallel database.

HadoopDB [Abouzeid et al., 2009] is one of the attempts at constructing such a hybrid system. It combines parallel databases and MapReduce to exploit both the high performance of the parallel database and the scalability of MapReduce. The basic idea behind HadoopDB is to use MapReduce as the communication layer above multiple nodes running single-node DBMS instances. Queries are expressed in SQL, translated into MapReduce by extending existing tools, and as much work as possible is pushed into the higher performing single node databases. In their experiments, they tested several frequently used SQL-queries, such as select query, join query, simple group-by query, etc. over one or more of the three relations.

Another way to realize such a hybrid solution is to integrate parallel database optimization as a part of calculations running with MapReduce. Since MapReduce does not give any limitations over the implementations, such a hybrid solution is totally feasible. Our work's approach also belongs to the hybrid solution.

5.5   SimMapReduce: Simulator   for   modeling   MapReduce framework

More attention is paid to MapReduce, not only by IT enterprises, but also by research institutes. The researchers make efforts on theoretical analysis on MapReduce computational model [Yang et al., 2007], scheduling mechanisms [Yu and Magoulès, 2007, Yu and Magoulès, 2008, Yu and Magoulès, 2009, Zaharia et al., 2008, Zaharia et al., 2009], task assignment [Ucar et al., 2006, Pan et al., 2010d], and workflow optimization, instead of implementing a real MapReduce application. In addition, different applications require different system configurations and parameters, so the construction of such real MapReduce systems is extremely challenging on a large scale of infrastructures. In view of the above considerations, simulation methods become a good alternative, which can accelerate study progress by opening the possibility of evaluating tests with hypothesis setting in advance and by simplifying the programming of implementation. It is easy to configure the infrastructures according to user requirements, and costs very little to repeatedly test various performances in a controllable manner.

Although there are some open source supporters of MapReduce implementation, few specific simulators exist to offer a simulated environment for MapReduce theoretical researchers. Therefore, a simulation tool, SimMapReduce, has been developed to simulate the performance of different applications and scenarios using MapReduce framework. The users of SimMapReduce only concentrate on specific research issues without getting concerned about finer implementation details for diverse service models.

5.5.1   Multi-layer architecture

A multi-layer architecture shown in Figure 5.4, is applied for the design of the SimMapReduce simulator for two reasons. The first is that layered design classes have the same module dependency. It is much clearer for both simulator designers and users than plane architecture. The second is that existing technologies and packages are easily leveraged into SimMapReduce as separate components, so the reusable codes can save designer time and energy in similar circumstances. More specifically, SimJava and GridSim packages are used as the base layers of a SimMapReduce simulator to provide the entities, communication, and task modeling capacity.

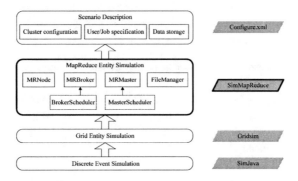

FIGURE 5.4: Four-layer architecture.

Discrete event simulation.   As a discrete event simulation infrastructure, SimJava consists of a collection of entities connected together by ports. The process of simulation advances through event delivery. Each entity responds to a coming event, and then sends the expected action to the next entity. The way of dealing with discrete events perfectly suits the simulation of the MapReduce framework, because entities are distributed in the cluster and Map/Reduce computations are sequential and parallel.

Grid entity simulation.   The GridSim toolkit supports entity modeling in distributed computing systems. It simulates geographically distributed resources in multiple administrative domains, and provides interfaces to fulfill resource management schemes. GridSim facilitates the basic provision of system components such as grid resource, broker, gridlet, workload trace, networks, and simulation calendar.

MapReduce entity simulation.   The higher level of simulation is the core of MapReduce functionality modeling, some of which is extended by the GridSim library. SimMapReduce toolkit can simulate various cluster environments regardless of small shared-memory machines, massively parallel supercomputers, or large collections of networked commodity PCs. Every node reserves separated slots for Map and Reduce. Broker takes the responsibility for allocating nodes to coming users. After a user receives a set of available nodes, the job dispatcher named master is in charge of mapping Map/Reduce tasks to a specific node. In the simulator, each job possesses one correspondent master. Although several traditional broker/master schedulers are integrated in SimMapReduce, advanced implementation of scheduling algorithms and policies is open to users. They are free to achieve multi-layer scheduling schemes on user-level and task-level. These algorithms can be conveniently overwritten on the basis of predefined abstract classes. Besides, the file transmission time is included in the completion time of jobs, which is monitored by a FileManager. FileManager can be considered as an abstract function entity of a HDFS Namenode, which manages the file system namespace and operations related to files, such as input files initiation, intermediate file management and file transmission.

Scenario description. The top layer is open for users of SimMapReduce. Different simulation scenarios are modeled by defining specific parameters in a configuration file in a quick manner, so that identical results are easily promised by repeated simulations. Extensive Markup Language (XML) is a set of rules for encoding documents in machine readable form. The design goals of XML emphasize simplicity, generality, and usability over the Internet. Many application programming interfaces (APIs) have been developed to help software developers process XML data. Therefore, a XML file is a good choice for the system configuration file of the simulator.

5.5.2   Input and output of simulator

In SimMapReduce, system parameters are configured in the file configure.xml, including three parts: cluster configuration, user/job specification, and data storage.

Cluster configuration. The cluster consists of a number of computing resources. Each resource, named node, encompasses several homogeneous or heterogeneous machines. The type of machine is predefined, varying the number of cores and the Millions of Instruction per Second (MIPS) rating. In order to monitor MapReduce node scheduling, each node reserves a certain number of slots for Mapper and Reducer, respectively. The active execution can not exceed the max slot limitation, if more than one task arrives. The computing capacity is scheduled by round robin algorithm, except that all tasks are executed at the same time. The network simulation is based on Gridsim. Routing information protocol is used by the router. Links introduce propagation delays, baud rate and the maximum transmission unit (MTU) to facilitate data transmission through a link. An example of cluster configuration is given by Figure 5.5.

```
<root>

<!--********************** Machine definition    ***********************-->
<Machine id="0" numPE="2" ratingPE="560" />
<Machine id="1" numPE="4" ratingPE="560" />

<!--********************** MachineList definition ***********************-->
<MachineList name="ml1"> 0,1 </MachineList>

<!--********************** Resource definition    ***********************-->
<!-- resCha is the resourceCharactristics,   ADVANCE_RESERVATION = 4,OTHER_POLICY_DIFFERENT_RATING=3,
     OTHER_POLICY_SAME_RATING=2 SPACE_SHARED=1,TIME_SHARED=0
-->
<Resource name="r1" baud_rate="1000" peakLoad="0.0" offPeakLoad="0.0" holidayLoad="0.0"
arch="Sun Ultra" os="Solaris" time_zone="9" cost="3.0" resCha="0" maxMpn="1" maxRpn="1"> ml1
</Resource>

<Resource name="r2" baud_rate="1000" peakLoad="0.0" offPeakLoad="0.0" holidayLoad="0.0"
arch="Sun Ultra" os="Solaris" time_zone="9" cost="3.0" resCha="0" maxMpn="1" maxRpn="1"> ml1
</Resource>

<Resource name="r3" baud_rate="1000" peakLoad="0.0" offPeakLoad="0.0" holidayLoad="0.0"
arch="Sun Ultra" os="Solaris" time_zone="9" cost="3.0" resCha="0" maxMpn="1" maxRpn="1"> ml1
</Resource>

<Resource name="r4" baud_rate="1000" peakLoad="0.0" offPeakLoad="0.0" holidayLoad="0.0"
arch="Sun Ultra" os="Solaris" time_zone="9" cost="3.0" resCha="0" maxMpn="1" maxRpn="1"> ml1
</Resource>

<!--********************** Router definition    ***********************-->
<Router routerName="Router1" routerClass="gridsim.net.RIPRouter"/>

<!--********************** Link definition    ***********************-->
<!-- baud_rate, bits/sec; propDelay, propagation delay in millisecond;
     mtu, max. transmission unit in byte -->
<Link linkName="link1" baud_rate="1000" propDelay="10" mtu="50" entity1="Router1" entity2="r1"/>
<Link linkName="link1" baud_rate="1000" propDelay="10" mtu="50" entity1="Router1" entity2="r2"/>
<Link linkName="link1" baud_rate="1000" propDelay="10" mtu="50" entity1="Router1" entity2="r3"/>
<Link linkName="link1" baud_rate="1000" propDelay="10" mtu="50" entity1="Router1" entity2="r4"/>

</root>
```

FIGURE 5.5: Example of cluster configuration.

```
<root>
<!--********************** Broker definition    ***********************-->
<Broker brokerName="Broker1" baud_rate="1000" nodeAllocPolicyClass="org.buaa.mrsimu.SimpleNodeAllocationPolicy">
 User1,User2,
</Broker>

<!--********************** User definition    ***********************-->
<!-- Here lambda is the job's period, We add also the beta (the ratio of Map/Reduce) and the execution time of job
which is considered to execute in a homogeneous platform
-->
<User userName="User1" id="1" userOrg="Beihang" baud_rate="1000" lambda="60" exec_time="1.125">
Job1, Job2, Job3
</User>

<User userName="User2" id="2" userOrg="Beihang" baud_rate="1000" lambda="120" exec_time="1.125">
Job4
</User>

<!--********************** Job definition    ***********************-->
<!-- ********************** User 1 ********************************************** -->
<Job jobName="Job1" masterName="Mas1" mapTaskLength="1000" reduceTaskLength="1000" inputFileSize="64"
intermediateFileSize="64" outputFileSize="64" numberOfMap="3" numberOfReduce="1" userName="User1"/>

<Job jobName="Job1" masterName="Mas1" mapTaskLength="1000" reduceTaskLength="1000" inputFileSize="64"
intermediateFileSize="64" outputFileSize="64" numberOfMap="3" numberOfReduce="1" userName="User1"/>

<Job jobName="Job1" masterName="Mas1" mapTaskLength="1000" reduceTaskLength="1000" inputFileSize="64"
intermediateFileSize="64" outputFileSize="64" numberOfMap="3" numberOfReduce="1" userName="User1"/>
<!-- ********************** User 2 ********************************************** -->
<Job jobName="Job4" masterName="Mas20" mapTaskLength="1000" reduceTaskLength="1000" inputFileSize="128"
intermediateFileSize="128" outputFileSize="128" numberOfMap="5" numberOfReduce="2" userName="User2"/>

<!--********************** Master definition    ***********************-->
<Master masterName="Mas1" jobName="Job1" baud_rate="1000" masterClass="org.buaa.mrsimu.SimpleMRMaster" />
<Master masterName="Mas2" jobName="Job2" baud_rate="1000" masterClass="org.buaa.mrsimu.SimpleMRMaster" />
<Master masterName="Mas3" jobName="Job3" baud_rate="1000" masterClass="org.buaa.mrsimu.SimpleMRMaster" />
<Master masterName="Mas4" jobName="Job4" baud_rate="1000" masterClass="org.buaa.mrsimu.SimpleMRMaster"/>
</root>
```

FIGURE 5.6: Example of user/job specification.

User/job specification.   Job stands for one MapReduce application running on a cluster. Each job consists of several Map and Reduce tasks. The task computation time is decided by the job length expressed in millions instruction (MI), not by input data. The input of Map task is the data stored on a cluster, and its size usually follows chunk splitting convention, 64M, for example. Intermediate file is considered as the output of Map task as well as the input of Reduce task. The size of output file depends on specific applications. For a sort job, the output size equals the input size. In comparison, the output size for a search application is much smaller, because the search result might just be a figure or a word.

Users submit jobs to a cluster through a broker. Jobs belonging to one user arrive simultaneously or in time sequence. Besides, the arrival rate is specified in advance; the user could also assign priorities to jobs according to their importance. An example of user/job specification is given by Figure 5.6. Initial data layout is about

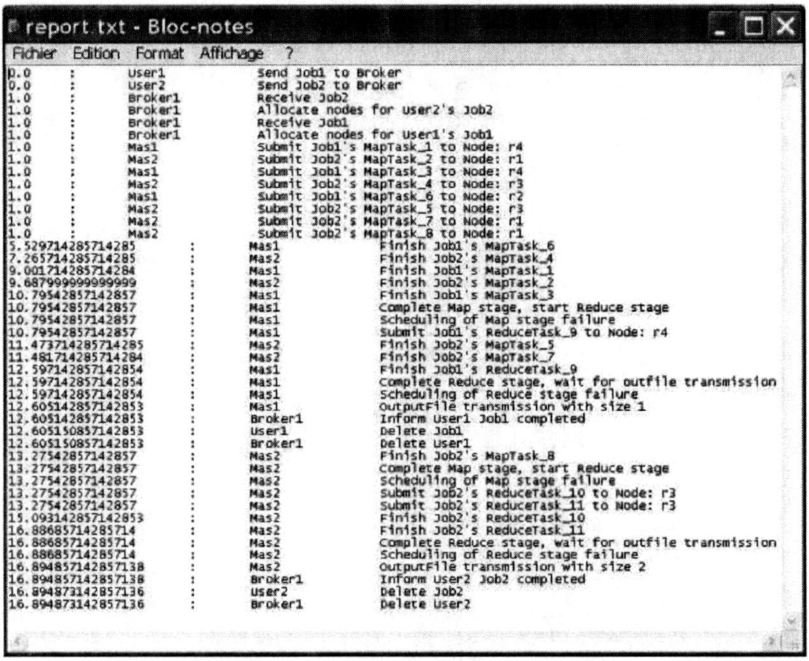


```
0.0                 :  User1      Send Job1 to Broker
0.0                 :  User2      Send Job2 to Broker
1.0                 :  Broker1    Receive Job2
1.0                 :  Broker1    Allocate nodes for User2's Job2
1.0                 :  Broker1    Receive Job1
1.0                 :  Broker1    Allocate nodes for User1's Job1
1.0                 :  Mas1       Submit Job1's MapTask_1 to Node: r4
1.0                 :  Mas2       Submit Job2's MapTask_2 to Node: r1
1.0                 :  Mas1       Submit Job1's MapTask_3 to Node: r4
1.0                 :  Mas2       Submit Job2's MapTask_4 to Node: r3
1.0                 :  Mas1       Submit Job1's MapTask_6 to Node: r2
1.0                 :  Mas2       Submit Job2's MapTask_5 to Node: r3
1.0                 :  Mas2       Submit Job2's MapTask_7 to Node: r1
1.0                 :  Mas2       Submit Job2's MapTask_8 to Node: r1
5.529714285714285   :  Mas1       Finish Job1's MapTask_6
7.265714285714285   :  Mas2       Finish Job2's MapTask_4
9.001714285714284   :  Mas1       Finish Job1's MapTask_1
9.687999999999999   :  Mas2       Finish Job2's MapTask_2
10.79542857142857   :  Mas1       Finish Job1's MapTask_3
10.79542857142857   :  Mas1       Complete Map stage, start Reduce stage
10.79542857142857   :  Mas1       Scheduling of Map stage failure
10.79542857142857   :  Mas1       Submit Job1's ReduceTask_9 to Node: r4
11.473714285714285  :  Mas2       Finish Job2's MapTask_5
11.481714285714284  :  Mas2       Finish Job2's MapTask_7
12.597142857142854  :  Mas1       Finish Job1's ReduceTask_9
12.597142857142854  :  Mas1       Complete Reduce stage, wait for outfile transmission
12.597142857142854  :  Mas1       Scheduling of Reduce stage failure
12.605142857142853  :  Mas1       OutputFile transmission with size 1
12.605142857142853  :  Broker1    Inform User1 Job1 completed
12.605150857142853  :  User1      Delete Job1
12.605150857142853  :  Broker1    Delete User1
13.27542857142857   :  Mas2       Finish Job2's MapTask_8
13.27542857142857   :  Mas2       Complete Map stage, start Reduce stage
13.27542857142857   :  Mas2       Scheduling of Map stage failure
13.27542857142857   :  Mas2       Submit Job2's ReduceTask_10 to Node: r3
13.27542857142857   :  Mas2       Submit Job2's ReduceTask_11 to Node: r3
15.093142857142853  :  Mas2       Finish Job2's ReduceTask_10
16.88685714285714   :  Mas2       Finish Job2's ReduceTask_11
16.88685714285714   :  Mas2       Complete Reduce stage, wait for outfile transmission
16.88685714285714   :  Mas2       Scheduling of Reduce stage failure
16.894857142857138  :  Mas2       OutputFile transmission with size 2
16.894857142857138  :  Broker1    Inform User2 Job2 completed
16.894873142857136  :  User2      Delete Job2
16.894873142857136  :  Broker1    Delete User2
```

FIGURE 5.7: Example of output.

the location of data chucks on a cluster. As the input file of Map tasks, data storage and transferring affects the computation performance for the Map phase, even for the overall job. We assume a uniform distribution as default. However, our design is flexible and other distributions are allowed for particular tests.

Output. The output of SimMapReduce is a report.txt, which provides a detailed execution trace. The trace can be shown in a coarse or fine manner. The former records phase-level time execution for jobs, while the latter is able to record every event. An example of simulator output is shown in Figure 5.7. Every row begins with the time, and is followed by the name of an entity and its behavior.

5.5.3 Implementation details of simulator

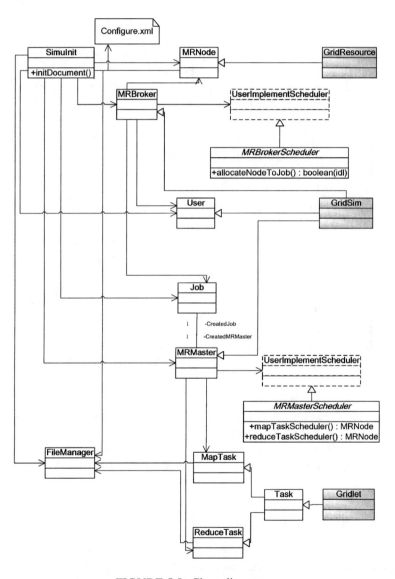

FIGURE 5.8: Class diagram.

The Class diagram is shown in Figure 5.8; the gray ones are parent classes archived by Gridsim.

MRNode.   This class models the computing infrastructure, each instance of which stands for a physical node on a cluster. Modelers can vary the characteristics such as processor number, speed and reserved Map/Reduce slot number. Input data is stored on disk within the given storage. Furthermore, this class is in charge of the receiving, executing, returning of submitted Map/Reduce task and input/intermediate/output file transferring.

MRBroker.   This class models the mediating broker between both sides of supply and demand. It is equipped with several lists of updated information about node, user and job, and it is capable of allocating proper nodes to jobs according to QoS needs. The concrete allocation policy must be pointed out in the MRBrokerScheduler.

MRBrokerScheduler.   This abstract class provides the possibility for modelers to designate the scheduling algorithm used by MRBroker. A modeler can integrate criteria such as client priority cost, deadline, due time, and flow to draw up a reasonable allocation policy. The default implementation is SimpleMRBrokerScheduler, which allocates all nodes of a cluster to every incoming job.

User.   This class models the resource demander, each instance of which represents a natural MapReduce client who communicates with the broker directly. It consists of a sequence of jobs that arrive simultaneously, randomly, or repeatedly. As in a real market, MapReduce users are assigned to ranks according to their priorities.

Job.   This class models the core functional MapReduce service, which is deployed on a group of nodes. It records every detail of service demands including arrival time, deadline, program operations, granularity, and quantity of Map/Reduce tasks, location and size of files.

MRMaster.   This class models the entity which takes responsibility for assigning and dispatching a Map/Reduce task to one node, managing intermediate files, buffering key/value pairs and supporting scheduling in static or dynamic manners. The concrete heuristic policy must be pointed out in MRMasterScheduler.

MRMasterScheduler.   This abstract class defines abstract methods (e.g. mapTaskScheduling and reduceTaskScheduling) which should be implemented by users. Several elements must be taken into account for the implementation of these abstract methods, such as data locality, interdependence between Map and Reduce, and processor throughput. A default SimpleMRMasterScheduler realizes strict local assignment for Map tasks and random assignment for Reduce tasks.

Task. This class models the finest unit for a MapReduce job. It can be subdivided into two types, MapTask and ReduceTask. After all the MapTasks finish, ReduceTasks are created by MRMaster depending on the location of intermediate key/-value pairs. The distinction between the two types of task mainly lies in the different input and output files.

FileManager. This class models a manager taking charge of all operations related to files, including recording, inquiring, tracing, updating and so on. This entity built on the fact that a typical MapReduce computation processes massive data files on an elastic cluster.

SimuInit. This class models initialization of simulation. It reads the parameter values into the instances of class and starts the simulation.

5.5.4   Modeling process

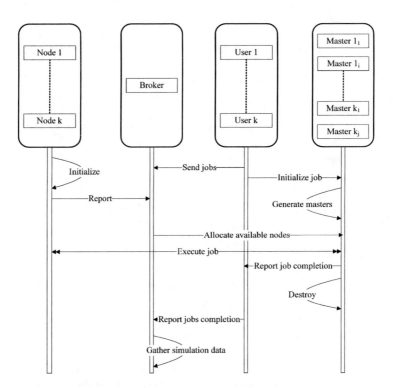

FIGURE 5.9: Communication among entities.

Since SimMapReduce is built on the discrete event simulation package SimJava, it contains a few entities running in parallel in their own threads. The entities represent physical objects in real MapReduce simulation, and create a network to communicate with each other by sending and receiving messages through SimJava's timestamp event queue.

Main entities for node, broker, user and master are implemented by separated classes discussed above. The communication among them is shown in Figure 5.9

In the beginning, nodes in a cluster report their characters to broker. At the same time, users initialize their own job sequences, and send jobs one by one, depending on arrival rate. An arbitrary job generates the amount of ordinary copies naming MapTask and ReduceTask, as well as a special copy of an operation program, the master, acting on behalf of job.

In every round, the master first sends information to the broker to request available nodes. MRBroker matches both sides' requirement and allocates a number of nodes to the master for its inner scheduling. The master manages the scheduling of Map/Reduce tasks, and supervises their execution. When all subtasks have been completed, the master reports job completion to the user and destroys itself. Concrete control flow of the master is shown in Figure 5.10.

When a user has completed all jobs in the sequence, it informs the broker of the information. If no more jobs are created, the broker gathers the simulation data and finishes simulation. The MRMaster is in charge of spawning Map and Reduce tasks, scheduling tasks to working nodes, managing their associate data, and producing the final output file. Every process is triggered by an event message. Having the available node list, MRMaster picks idle nodes to schedule MapTasks. As soon as a node receives a MapTask, it checks whether the input file is on local disk. If not, the node asks for input transmission. When input data is ready, MapTask runs its Map function. After that, intermediate files produced by Map operations are buffered on memory. MapTask then reports its completion to MRMaster. MRMaster keeps on examining whether all MapTasks finish. If yes, MRMaster stops the Map phase, and starts the shuffle phase that groups the key/value pairs by common values of the key. Generally, data with the same key will be sent to one ReduceTask.

In the begining of Reduce phase, MRMaster makes a scheduling decision to dispatch ReduceTasks to different nodes. The first action in the Reduce phase is reading intermediate files remotely. In our current design, each ReduceTask receives an equal part from each MapTask output. Thus the input of ReduceTask sums up all intermediate files regardless of weight. Then the Reduce function is operated, generating output file. Similarly as in the Map phase, the MRMaster collects the message about the completion of the ReduceTask. When they all finish, a final output result is obtained. The computation of the job terminates, and its manager, the MRMaster breaks down.

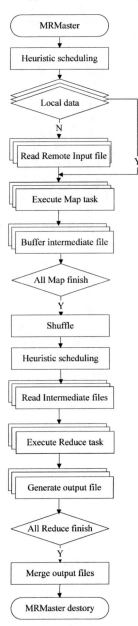

FIGURE 5.10: Control flow of MRMaster.

5.6 Concluding remarks

In this chapter, we first introduced the basic idea and related issues of MapReduce model and its extended model, MapCombineReduce. Two implementation frameworks of MapReduce, Hadoop and GridGain were presented. They have different latency. Hadoop has high latency, while GridGain has low latency. Under the interactive response time requirement, GridGain is a suitable choice for our work. The MapReduce model hides the underlying communication details. We analysed, in particular, the communication cost of the MapReduce procedure, and discussed the main factors that influence the communication cost. We then discussed the job-scheduling issues in MapReduce. In MapReduce job-scheduling, two more things need to be considered than in other cases: data locality, and dependence between *mapper* and *reducer*. Our discussion also involves MapReduce efficiency and its application on different hardware. Second, we described the distributed data storage underlying MapReduce, including distributed filesystems, like GFS and its open source implementation—HDFS, an efficient enhanced storage system based on a cache mechanism. Another approach is manual support of MapReduce data access adopted in our work. The third topic addressed in this chapter is data management in the cloud. The suitability of being processed with MapReduce was discussed for transactional data management and analytical data management. The latter was thought to be able to benefit from MapReduce model. Relying on this, we further addressed large-scale data analysis based on MapReduce. We presented the MapReduce-based data query languages and data analysis related work with MapReduce. As shared-nothing parallel database and MapReduce system use similar hardware, we focused on comparing them and followed by presenting the related work on a hybrid solution combining these two into one system. Finally, we introduced related parallel computing frameworks.

Chapter 6

Large-scale multi-dimensional data aggregation

6.1 Introduction

In this chapter, we will present the MapReduce-based multi-dimensional data aggregation. We will first describe the background of our work, as well as the organization of data used in our work. Then we will introduce *Multiple Group-by* query, which is also the calculation that we will parallelize relying on MapReduce. We will give two implementations of *Multiple Group-by* query, one is based on MapReduce, and the other is based on MapCombineReduce. The job definitions for each implementation will be specifically described. We will also present the performance measurement and execution time estimation.

6.2 Data organization

The traditional data cube is stored either in the form of multi-dimensional data array (in MOLAP) or under star-schema (in ROLAP). MOLAP suffers from sparsity of data. When the number of dimensions increases, the sparsity of the cube also increases at a rapid rate. Sparsity is an insurmountable obstacle in MOLAP. In addition, MOLAP pre-computes all the aggregates. When the amount of data is large enough, pre-computation will take a long time. Thus, MOLAP is only suitable for datasets of small and moderate size. In contrast, ROLAP is more suitable for datasets of large size. In ROLAP, one of the traditional query accelerating approaches is pre-computing[1]. The pre-computing approach requires that all the aggregated values contributing to the potential queries be computed before processing queries. For this purpose, the database administrators need to identify the frequently demanded queries from numerous past queries, and then build materialized views and indexes for these queries. Figure 6.1 shows the data organization in case of employing materialized views. Certainly, query response time is reduced by this approach. However,

---

[1]Refer to Section 4.2 for more information.

the computations involved in this approach are heavy; more calculations are needed for choosing the optimal one from multiple materialized views during query processing. Materialized views can only help to accelerate processing of a certain set of pre-chosen queries, not to accelerate the processing of all queries. Another disadvantage of materialized views is that they take up a lot of storage space.

An alternative approach to organizing data is to store one overall materialized view. The materialized view is a result of join operations among all dimension tables and the fact table. Assume a dataset is from an on-line apparel selling system, recording the sales records of all products in different stores during the last three years. This dataset is originally composed of four dimension tables (COLOR, PRODUCT, STORE, WEEKS) and one fact table (FACT). Refer to the sub-figure (a) of Figure 6.1 for the star-schema definition of the dataset. The overall materialized view is named ROWSET. Each record stored in ROWSET contains values of all measures retrieved from the original FACT table associated with the distinct values of different dimensions retrieved from dimension tables. The List 6.1 shows the SQL statements used to generate the materialized view, ROWSET. Those SQL statements are written in *psql*, which can be interpreted and executed in PostgreSQL. A graphic illustration is available in Figure 6.2:

Listing 6.1: SQL statements used to create materialized view—ROWSET

```sql
DROP TABLE IF EXISTS "ROWSET" CASCADE;
CREATE TABLE "ROWSET" AS
SELECT c."color_name" AS color,
       p."family_name" AS product_family,
       p."family_code" AS product_code,
       p."article_label" AS article_label,
       p."category" AS product_category,
       s."store_name" AS store_name,
       s."store_city" AS store_city,
       s."store_country" AS store_state,
       s."opening_year" AS opening_year,
       w."week" AS week,
       w."month" AS "month",
       w."quarter" AS quarter,
       w."year" AS "year",
       f."quantity_sold" AS quantity_sold,
       f."revenue" AS revenue
FROM   "WEEKS" w JOIN
       ("COLOR" c JOIN
       ("PRODUCT" p JOIN
       ("STORE" s JOIN
        "FACT" f ON (f."store_id"=s."store_id"))
        ON (f."product_id"=p."product_id"))
        ON (f."color_id"=c."color_id"))
            ON (w."week_id"=f."week_id");
```

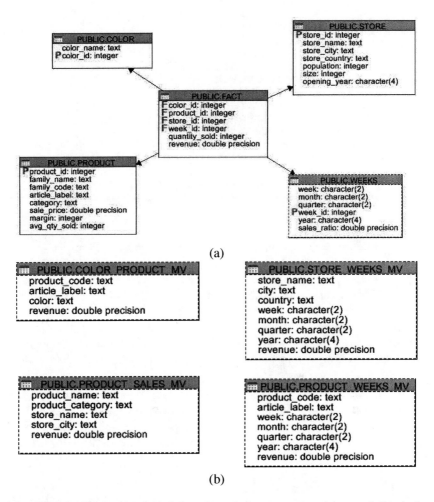

(a)

(b)

FIGURE 6.1: Storage of original dataset and the pre-computed materialized views for the identified, frequently demanded queries: sub-figure (a) shows original data set under star-schema; sub-figure (b) shows several materialized views created based on the original dataset represented in (a).

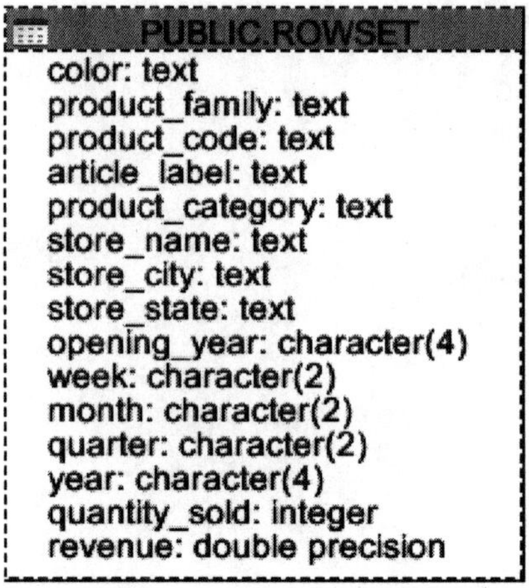

```
╔══════════════════════════════════╗
║ ▦▦▦▦▦▦  PUBLIC.ROWSET  ▦▦▦▦▦▦▦▦ ║
╟──────────────────────────────────╢
│ color: text                      │
│ product_family: text             │
│ product_code: text               │
│ article_label: text              │
│ product_category: text           │
│ store_name: text                 │
│ store_city: text                 │
│ store_state: text                │
│ opening_year: character(4)       │
│ week: character(2)               │
│ month: character(2)              │
│ quarter: character(2)            │
│ year: character(4)               │
│ quantity_sold: integer           │
│ revenue: double precision        │
└──────────────────────────────────┘
```

FIGURE 6.2: Overall materialized view—*ROWSET*.

In this work, the data organization of a single overall materialized view is adopted. It has several advantages. First, compared with multiple materialized views approaches, the required storage space is reduced. Second, the overall materialized view is not created for optimizing a set of pre-chosen queries, instead, all queries can benefit from this materialized view. Third, with one materialized view, the emerging search engine techniques could be easily applied. In particular, in this work, one materialized view approach greatly simplifies data partitioning and indexing work.

Term specification.   From now on, the terms *dimension* and *measure* are slightly different from the recognized terms with common names in the OLAP field. In order to avoid confusion, we would like to newly declare the definitions of these two terms. If not additionally specified, the following occurrences of the two terms adopt the definitions below.

- *Dimension* is a type of column, of which the distinct values are of type *text*.

- *Measure* is a type of column, of which the distinct values are of type *numeric*.

To be noted, *hierarchy* is not adopted in this terminology.

6.2.1   Computations in data explorations

Usually in data explorer products, the user selects an information space, and then enters into a relevant exploration panel. The first page displayed in the exploration panel shows aggregated measures dimension by dimension. The user can select various aggregate functions, such as COUNT, SUM, AVERAGE, MAX, MIN, etc, by clicking on a drop-list. Assuming the aggregate function SUM applied on all the measures, then the computations involved within the display of the first page are actually equivalent to execution of SQL statement of the List 6.2.

Listing 6.2: SQL statements used for displaying the first page of exploration panel.

```
DROP VIEW IF EXISTS page_0 CASCADE;
CREATE VIEW page_0 AS
SELECT * FROM "ROWSET"
;
DROP VIEW IF EXISTS dimension_0;
CREATE VIEW dimension_0 AS
SELECT "page_0"."color" AS distinct_value,
       SUM(page_0."quantity_sold") AS quantity_sold,
       SUM(page_0."revenue") AS revenue
FROM "page_0"
GROUP BY "distinct_value"
;
DROP VIEW IF EXISTS dimension_1;
CREATE VIEW dimension_1 AS
SELECT "page_0"."product_family" AS distinct_value,
       SUM(page_0."quantity_sold") AS quantity_sold,
       SUM(page_0."revenue") AS revenue
FROM "page_0"
GROUP BY "distinct_value"
;
DROP VIEW IF EXISTS dimension_2;
CREATE VIEW dimension_2 AS
SELECT "page_0"."product_category" AS distinct_value,
       SUM(page_0."quantity_sold") AS quantity_sold,
       SUM(page_0."revenue") AS revenue
FROM "page_0"
GROUP BY "distinct_value"
;
——————————repeat for all the dimensions——————
```

If a user finds an anomalous aggregated value, for example, a certain product category, say, "swimming hats" has too low a quantity sold, and he/she wants to see the detail data over the specific product category, then a detailed exploration is performed and the second page is generated. The second page displays the "swimming hats" related measures aggregated over different dimensions. The computation involved in displaying the second page of exploration panel is equivalent to execution of SQL statement in List 6.3.

Listing 6.3: SQL statements used for displaying the second page of exploration panel.

```
DROP VIEW IF EXISTS page_1 CASCADE;
CREATE VIEW page_1
AS
SELECT * FROM "page_0"
WHERE "product_category"='Swimming_hats'
;
DROP VIEW IF EXISTS dimension_0;
CREATE VIEW dimension_0 AS
SELECT "page_1"."color" AS distinct_value ,
       SUM( page_1."quantity_sold") AS quantity_sold ,
       SUM( page_1."revenue") AS revenue
FROM "page_1"
GROUP BY "distinct_value"
;
DROP VIEW IF EXISTS dimension_1;
CREATE VIEW dimension_1 AS
SELECT "page_1"."product_family" AS distinct_value ,
       SUM( page_1."quantity_sold") AS quantity_sold ,
       SUM( page_1."revenue") AS revenue
FROM "page_1"
GROUP BY "distinct_value"
;
DROP VIEW IF EXISTS dimension_2;
CREATE VIEW dimension_2 AS
SELECT "page_1"."product_category" AS distinct_value ,
       SUM( page_1."quantity_sold") AS quantity_sold ,
       SUM( page_1."revenue") AS revenue
FROM "page_1"
GROUP BY "distinct_value"
;
────────────repeat for all the dimensions────────────
```

Similarly, further exploration can be achieved by applying both the current *WHERE* condition `"product_category"="Swimming hats"` and the new condition coming from the exploration panel.

As seen from the above illustration, a typical computation involved in data exploration is the *Group-by* query, on different dimensions, with aggregates using various aggregate functions. Without considering other features of data explorer, we could say that the computations involved in a data exploration are composed of a couple of elementary *Group-by* queries having the following form:

```
SELECT DimensionA ,
       aggregate_funtion(Measure1),
       aggregate_funtion(Measure2)
FROM "ROWSET"
```

```
WHERE DimensionB=b
GROUP BY DimensionA ;
```

Group-by query is a typical OLAP query. Because of OLAP queries wide application, a lot of research work has been done. The characteristics of these queries are summarized in [Liao and Pei, 2008]. The two important characteristics of an OLAP query—including *Group-by* query—are:

- most of them include aggregate functions;

- they usually include selection clauses.

Going any further from the two characteristics, one *Group-by* query involves two processing phases, *filtering* and *aggregating*. During the filtering phase, the WHERE condition is applied to filter the records of a materialized view. During the aggregating phase, the aggregate function is performed over the filtered records.

6.2.2 Multiple group-by query

To be able to respond to a user's exploration, multiple *Group-by* queries need to be calculated simultaneously, instead of one single *Group-by* query. The term *Multiple Group-by query* expresses more clearly the characteristics of the query addressed in this work. Defining the *Multiple Group-by* query, we describe it as follows: To define *Multiple Group-By* query is a set of Group-by queries using the same select-where clause block. More formally, a *Multiple Group-By* query can be expressed in SQL in the following form:

```
SELECT X, SUM(∗) ,
FROM R WHERE condition
GROUP BY X
ORDER BY X;
```

where X is a set of columns on relation R.

Some commercial database systems support a similar *Group-by* construct named *GROUPING SETS*, and it allows the computation of multiple Group-by queries using a single SQL statement [Zhimin and Vivek, 2005]. Compared with the *Multiple Group by* query addressed in this work, a GROUPING SETS query is slightly different. Each Group-by query contained in a GROUPING SETS query could have more than one group-by dimension, i.e. one Group-by query aggregates over more than one dimension, whereas in this work, one Group-by query aggregates over only one dimension.

In data exploration environment, processing *Multiple Group-by* query has several challenges. The first challenge is large data volume. In a very common case, the historical dataset is often of large size. The generated materialized view is also of large size. In order for a user to do analysis as comprehensively as possible, the historical dataset contains many dimensions. It is not rare that the generated overall materialized view has more than ten dimensions. The second challenge is the requirement of short response time. It is a common demand for all the interactive interface applications, including data exploration. *Multiple Group-by* queries aggregating over all dimensions is repeatedly invoked and processed during data exploration, then it is required that each query be answered within a very short time, for example, not more than five seconds, ideally; within hundredths of milliseconds. Summarizing these challenges' description—doing time and resource consuming computations in a short time.

Parallelization is the solution to address these challenges: partition the large materialized view ROWSET into smaller blocks, then process a query over each of them, finally merge the results. One particularity of this work is that we utilize cheap commodity hardware instead of expensive supercomputers. This is also the significant side of this work. This particularity brings further challenges, scalability and fault tolerance issues. To address these challenges, we adopt the MapReduce model, and the detailed specifications will be given later in this chapter.

6.3   Choosing a right MapReduce framework

There are several projects and research works focusing on building specific MapReduce frameworks for various hardware and different distributed architectures. In our work, we adopt the shared-nothing clusters, which are available for free[2]. Some well-designed MapReduce frameworks have already been realized for this type of hardware architecture. We need to choose the right framework to satisfy the specific requirement.

6.3.1   Advantages of GridGain

Interactivity, i.e. short response time, is the basic requirement in this work. In order to meet this requirement, while the application-level optimization is essential, choosing a right underlying MapReduce framework is also important. At the time we chose our framework, there were two different open-source MapReduce frameworks available, Hadoop [Hadoop, 2012a] and GridGain [GridGain, 2012].

At first, Hadoop was chosen as the MapReduce supporting framework. We successfully installed and configured Hadoop in a cluster of two computers, and ran several simple tests over Hadoop. The experiment execution time over Hadoop was not satisfying. An application of filtering materialized view of small size with a given condition already took several seconds, which is too slow for interactive interface. This phenomenon was then diagnosed as a consequence of the high latency of Hadoop. Actually, high latency is consistent with the initial design of Hadoop. Hadoop is designed to address batch-processing application. Batch-processing appli-

---

[2]Grid'5000, for more information, refer to [Grid'5000, 2012]

cation only emphasizes high-throughput. In such a context, high-latency is insignificant. The high-latency is also a side effect of Hadoop's "MapReduce + HDFS" design, of which more explanation can be found in Subsection 5.2.2.

Another MapReduce framework, GridGain, was finally adopted as the underlying framework in this work. GridGain offers a low latency since it is a pure MapReduce engine without being associated with a distributed file system. Therefore, data partitioning and distributing should be managed manually. Although this increases the workload of programmers, they have a chance to do optimizations at the data access level. Additionally, GridGain provides several pre-defined scheduling policies including data affinity scheduling policy, which can be beneficial for processing multiple continuous queries.

6.3.2   Combiner support in Hadoop and GridGain

The **Combiner** is an optional component, which is located between the **mapper** and the **reducer**. In Hadoop, the combiner component is implemented, and a user can choose to use or not to use it freely. The combiner is physically located on each computing node. Its function is to locally collect the intermediate outputs from the mappers running on the current node before these intermediate outputs are sent over the network. In certain cases, using this combiner component can optimize the performance of the entire model. The objective of using a combiner is to reduce the intermediate data transfer.

This optional component **combiner** is not implemented in GridGain. In this work, we propose a bypass method to make GridGain a supporting combiner. Although this method is implemented on top of GridGain, it is not limited to work with GridGain. The same idea can also be carried out on other MapReduce frameworks. We illustrate this method in the Figure 6.3. A GridGain MapReduce is composed of multiple mappers and one reducer. In this method, we utilized two successive GridGain MapReduce tasks. In the first MapReduce task, the mappers correspond to the mapper component of MapCombineReduce model, and its reducer acts as a trigger to activate the second MapReduce task, once the first MapReduce mappers have all finished their work. The mappers of the second MapReduce task actually act as the combiner component of MapCombineReduce model. The reducer of the second MapReduce task does the job of the reducer component of MapCombineReduce model.

6.3.3   Realizing MapReduce applications with GridGain

GridGain provides developers with Java-based technologies to develop and to run grid applications on private or public clouds. In order to implement a MapReduce application, there are mainly two classes needed to define in GridGain, i.e. **Task**, and **Job**. Task class definition requires a developer to realize `map()` and `reduce()` functions. Job class definition requires a developer to realize `execute()` function. The name of `map()` might be confusing. This name misleads people to think it defines the calculations to be carried out in a mapper. But in fact, the `map()` function

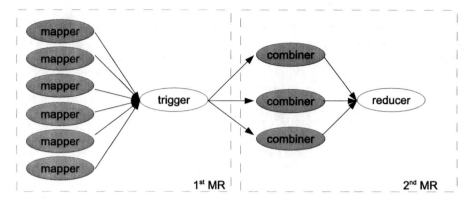

FIGURE 6.3: Creating the task of MapCombineReduce model by combining two GridGain MapReduce tasks.

is responsible for establishing the mappings between the mappers and the worker nodes. This function could be utilized to apply user-defined job-scheduling policy. GridGain provides some pre-defined implementations of map() function, which support various job-scheduling policies, including data affinity, round robin, weighted random, etc. The web site of GridGain [GridGain, 2012] gives the readers more descriptions of GridGain's supported job-scheduling policies. execute() is actually the function specifying operations performed by mapper. execute() contains the distributed computations which will be executed in parallel with other instances of the *execute()* function. When a mapper arrives at a remote worker node, a collision resolving strategy will look into a queue of existing mappers on this worker node to either reject the current mapper or leave it waiting in the queue. When the mapper runs, the execute() function will be executed. The Reduce function contains the actions of reducing, collecting mappers intermediate outputs and calculating the final result. It is usually composed of some aggregate-type operations. The Reduce() function's execution is activated by the arrival of the mappers' intermediate outputs. According to the policy defined by the user, reduce() can be activated once the first intermediate output from a mapper arrives at the master node, or after the sub-results of all the mappers have arrived. The default policy is to wait for all the mappers to finish their work and then to activate the reduce() method. In addition, the developer also needs to define a task loader program, which takes charge of initializing the parameters, starting a grid instance, launching the user's application, and then waiting for and collecting the results.

6.3.4 Workflow analysis of GridGain procedure

GridGain's MapReduce is composed of multiple mappers and one reducer. The mappers are sent to and run on worker nodes, and the reducer runs on the master node. To fully understand the procedure of GridGain's MapReduce, we analyzed the log file of GridGain and also did the profiling work when running the MapReduce application. The following description of GridGain MapReduce workflow is based on this analysis.

The Figure 6.4 shows the workflow of a GridGain's MapReduce task. When a master starts a MapReduce task, it in fact starts a thread for this task. The thread does the start-up work and closure work for the task. The start-up work includes the following steps.

1. First, the master creates mappings between user-defined jobs (mappers) and available worker nodes;

2. Second, the master serializes mappers in a sequential way ;

3. Once all the mappers are serialized, the master sends each mapper to the corresponding worker node.

After all the mappers have been sent, the thread terminates and the master enters into a "waiting" status. This "waiting" status continues until the master node receives the mappers' intermediate outputs. When the master node receives a mapper's intermediate output, it begins to de-serialize this intermediate output immediately. After all the intermediate outputs have been de-serialized, it then starts the reducer. The de-serialization and reducer execution compose the task's closure.

On the other side, the worker node listens to the messages after the GridGain instance is started. When it receives a message containing a serialized mapper object, it will de-serialize the message, and thereby obtain the mapper's object. Then, the mapper is put into a queue waiting to be executed once one or more CPU becomes available. After the mapper's execution is accomplished, the intermediate output is serialized and sent back to the master node.

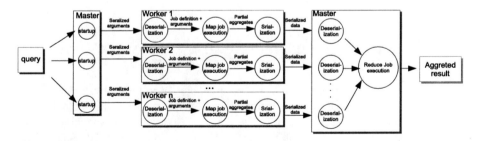

FIGURE 6.4: Workflow diagram of MapReduce-based application in GridGain.

6.4   Parallelizing single group-by query with MapReduce

Before addressing the parallelization of a *Multiple Group-by* query, we describe the processing of the elementary query—a single Group-by query—in MapReduce. Intuitively, a single Group-by query could be well matched with a MapReduce model. A single Group-by query can be executed in two phases: the first phase is filtering, and the second phase is aggregating. The other operations (regroup) can be incorporated into the aggregating phase. The filtering phase corresponds to the mappers' work in MapReduce, and the aggregating phase corresponds to its reducer's work.

As   an   example   of   Group-by   query,   we   consider   another materialized   view,   LINEITEM   with   two   dimensions   and   one   measure, LINEITEM(OrderKey,SuppKey,Quantity), and one Group-by query of the form:

```
SELECT "Orderkey", SUM("Quantity")
FROM "LINEITEM"
WHERE "Suppkey" = '4633'
GROUP BY Orderkey
```

The above query performs the following operations on the materialized view, LINEITEM. The first operation is **filtering**, which makes records to be filtered by the WHERE condition. Only the records matching the condition ""Suppkey" = '4633'" are retained for the subsequent operations. Within the next operation, these tuples are **regrouped** into groups according to the distinct values stored in the dimension OrderKey. The last operation is a SUM aggregation, which adds up the values of the measure Quantity. The SUM aggregation is executed on each group of tuples. Figure 6.5 illustrates how this MapReduce model-based processing procedure is organized.

6.5   Parallelizing multiple group-by query with MapReduce

A *Multiple Group-by* query can also be implemented in these two phases. In a *Multiple Group-by* query, with multiple single Group-by queries having the same WHERE condition, we propose that the mapping phase perform the computation for filtering data according to the condition defined by the common WHERE clause. The aggregating phase still corresponds to a set of reduce-type operations. For a *Multiple Group-by* query, the aggregating phase consists of a couple of aggregating operations performed on several different Group-by dimensions. In this work, we use the reducer to implement the aggregating phase at first, and then we propose an optimized

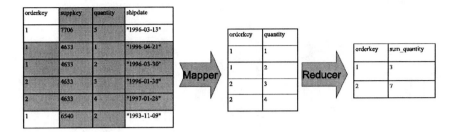

FIGURE 6.5: Single Group-by query's MapReduce implementation design. This design corresponds to the SQL query `SELECT Orderkey SUM(Quantity) FROM LINEITEM WHERE Suppkey = 4633 GROUP BY Orderkey`.

implementation based on the extended MapCombineReduce model. The following content in this section will give more details about these two implementations.

6.5.1 Data partitioning and data placement

The materialized view ROWSET used in our tests is composed of 15 columns, including 13 dimensions and 2 measures. We partition this materialized view into several blocks. The horizontal partitioning method [Stephano et al., 1982] is used to equally divide ROWSET. As a result, each block has an equal number of records, and each record keeps all the columns from the original ROWSET. All the data blocks are replicated on every participating worker node. This is inspired by the Adaptive virtual partitioning method proposed in [Lima et al., 2004a]. Such a method allows to conveniently realize the distribution of data without worrying about the accessibility problem caused by a data placement strategy. With all the data blocks available on all the worker nodes, the data location work is simplified.

6.5.2 MapReduce model-based implementation

The initial implementation of the MapReduce model-based *Multiple Group-by* query we have developed is shown in Figure 6.6. In this implementation, the mappers perform the filtering phase, and the reducer performs aggregating phase. In order to realize the filtering operations, each mapper first opens and scans a certain data block file, locally stored on the worker node, and then selects the records which meet the conditions defined in the WHERE clause. In this way, each mapper filters out a group of records. After that, all the records filtered by the mappers are sent to the reducer as intermediate outputs. The Algorithm 6.5.1 describes this processing with pseudo-code.

The reducer realizes the aggregating operations as follows. First, the reducer creates a set of aggregate tables to save the aggregate results. Each aggregate table

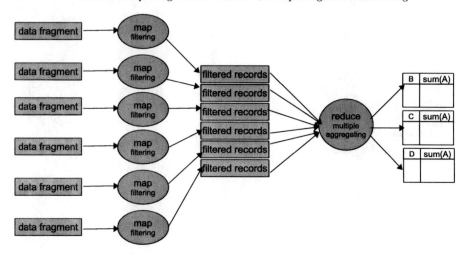

FIGURE 6.6: Initial *Multiple Group-by* query implementation based on the MapReduce model.

DV of Dimension	agg_func_1 (msr_x)	agg_func_2 (msr_y)	...
DV_1	agg_val_DV_1	agg_val_DV_1	...
DV_2	agg_val_DV_2	agg_val_DV_2	...
...	...	...	...

FIGURE 6.7: Aggregate table structure.

corresponds to a Group-by dimension. The **aggregate table** is a structure designed to store the aggregate value for one dimension. In addition to the distinct values of the dimension, the aggregate table also stores aggregate values calculated by applying user defined aggregate functions over different measures. As shown in Figure 6.7, the first column stores the distinct values of dimension, and the corresponding aggregate values are stored in the remaining columns. The number of aggregate functions (denoted as nb_{agg}) contained in the query determines the total number of columns. There are in total $nb_{agg} + 1$ columns in the aggregate table.

As an example, we specify the construction of **aggregate tables** for the *Multiple Group-by* query below:

```
SELECT
  SUM("revenue"), SUM("quantity_sold"), "product_family"
FROM "ROWSET"
WHERE "color"='Pink'
```

```
GROUP BY "product_family"
;
SELECT
  SUM("revenue"), SUM("quantity_sold"), "store_name"
FROM "ROWSET"
WHERE "color"='Pink'
GROUP BY "store_name"
;
SELECT
  SUM("revenue"), SUM("quantity_sold"), "year"
FROM "ROWSET"
WHERE "color"='Pink'
GROUP BY "year"
;
```

This *Multiple Group-by* query includes three single Group-by queries; each query includes two aggregate functions. Thus, we need to create three *aggregate tables*. For the first Group-by query aggregating over dimension product_family, the *aggregate table* has three columns. The first column is used to store different distinct values appearing in the records which meet the WHERE condition $E > e$. The second and third columns are used to store the corresponding aggregate values for each distinct value of the dimension product_family. In this example, the two aggregate functions are both SUM. The *aggregate table* for the second and third Group-by queries, are constructed in a similar way. The *aggregate table* is implemented as a Hash table in over program.

Second, the reducer scans all intermediate results, and simultaneously the reducer updates the **aggregate tables** by aggregating the newly arriving aggregate values onto some records in the **aggregate tables**. The final result obtained by the reducer is a group of aggregate result tables, each table corresponding to one Group-by query. The Algorithm 6.5.2 describes this processing in pseudo-code.

Algorithm 6.5.1 Filtering in Mapper.

Input: data block, *Multiple Group-by* query
Output *selectedRowSet*
Load data block into *rawData*
for $record \in rawData$ **do**
   if *record* passes WHERE condition **then**
      $recordID \rightarrow matchedRecordIDs$
   end if
end for
for $recordID \in matchedRecordIDs$ **do**
   copy $rawData[recordID]$ to *selectedRowSet*
end for

Algorithm 6.5.2 Aggregating in Reducer.

Input: $selectedRowSet$

Output: $aggs$

for $dimension \in GroupByDimensions$ **do**

    create a aggregate table: agg

end for

for $record \in selectedRowSet$ **do**

    for $dimension \in GroupByDimensions$ **do**

        if (value of $dimension$ in $record$) $\in agg$ of $dimension$ **then**

            assuming existing record is r

            for $agg\_func() \in agg\_func\_list$ **do**

                update $r.field(1+a)$ with $agg\_func(r)$

            end for

        else

            Insert into agg a new record rr where

            $rr.field1$ = value of $GroupByDimension$

            for $agg\_func() \in agg\_func\_list$ **do**

                $rr.field(1+a) = agg\_func(rr)$

            end for

        end if

    end for

end for

In this initial implementation, the reducer works on all the records filtered by the WHERE condition. The most important calculations, i.e. the aggregations, are performed in the reducing phase. It takes all the filtered records as its input data. Such an implementation is a general approach to realizing a MapReduce application. However, it is not fully suitable for GridGain. Because of the limitation of GridGain (only one reducer), all the filtered records should be transferred over the network. This could cause high overhead when the bandwidth is limited.

6.5.3   MapCombineReduce model-based implementation

In the initial implementation, all the intermediate outputs produced by the mappers (i.e. all the records matching the WHERE condition), are sent to the reducer over the network. If query selectivity [3] under the given WHERE condition is relatively small, for instance 1%[4], then output of mapping phase will be moderate, and the initial implementation is suitable. However, if the query selectivity is larger, for instance, 9%, then the number of records will be great and the volume of data being

---

[3]Here, a selectivity of a select query means the ratio value between the number of records satisfying the predicate defined in the WHERE clause and the cardinality of the relation.

[4]This means that only 1% of the records are selected from the data source table.

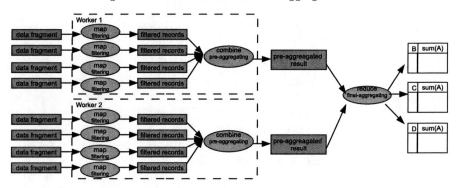

FIGURE 6.8: Optimized *Multiple Group-by* query implementation based on Map-CombineReduce model.

transferred over the network will become large, which results in a higher communication cost. As a consequence, the initial implementation is not suitable for queries with relatively large selectivity.

In order to reduce the network overhead caused by the intermediate data transmission for queries with larger selectivity value, we propose a MapCombineReduce model-based implementation. We let the combiner component act as a preaggregator on each worker node. In this work, the number of combiners of each worker node is one. In optimized MapCombineReduce model-based implementation, the mapper first performs the same operations of the filtering phase as in the initial implementation. However, the result of the filtering phase will be put into the local cache instead of being sent over the network immediately. The mapper then sends out a signal when it has finished its work. The trigger (i.e. reducer in the first MapReduce task) will receive this signal. When the trigger receives all the *work finished* signals, then it activates the second MapReduce task. In the second MapReduce task, the combiner (i.e. the mapper of the second MapReduce task) does the aggregating operations locally within each worker node. Each of the combiners generates a portion of aggregate the results, i.e. a set of partial aggregate tables. Then they send out their partial results to the reducer. After merging all the partial results, the reducer generates the final aggregate tables of the *Multiple Group-by* query. Thus, the volume of data to be transferred is reduced during the pre-aggregation phase, which in turn reduces the total communication cost. Figure 6.8 illustrates the MapCombineReduce model-based *Multiple Group-by* query processing.

6.6   Cost estimation

In this section, we will describe a basic estimation of execution time of the *Multiple Group-by* query. This cost estimation respectively addresses the initial implementation based on the MapReduce and optimized implementation based on the MapCombineReduce. As mentioned earlier, a GridGain MapReduce task calculation is composed of the start-up and closure on the master node, and the mapper executions over worker nodes. In this work, we are also interested in the optimization over communication cost, thus, communication time is considered, but we ignore the extra cost incurred by the resource contention over each worker node when running multiple mappers.

6.6.1   MapReduce model-based implementation

Assume that the Multiple Group-by query runs over a materialized view of N records, and the query has nb_{GB} Group-by dimensions. We use MapReduce-based method to parallelize query processing. The parallel *Multiple Group-by* query's total cost is composed of four parts:

- Start-up cost(on master node), denoted as C_{st};

- Mapper's execution (on workers), denoted as C_m;

- Closure cost (on master node), denoted as C_{cl};

- Communication cost, denoted as C_{cmm}.

In start-up, the master does the preparation of mappers, including the mappings from mappers to available worker nodes, and then sequentially performs the serializations of mappers with their attached arguments. We use C_{mpg} to denote the time for building mapping between mappers and worker nodes, C_s the time for serializing one unit size of data, $size_m$ the size of mapper object, nb_m the number of mappers. The notations that are used for expressing the cost estimation are listed in the Table 6.1.

Without considering the low-level details of serialization[5], we simply assume that the serialization time is proportional to the size of data being serialized, i.e. a bigger mapper object consumes more time during the serialization. Therefore, a MapReduce task of GridGain, the start-up cost C_{st} over master node is:

$$C_{st} = (C_{mpg} + C_s \cdot size_m) \cdot nb_m$$

When a worker node receives a message containing a mapper, it first serializes the mapper as well as its arguments; it then launches the execution of the mapper. When

---

[5]The details of serialization will be discussed in Chapter 7.

Table 6.1: Notations Used for Describing the Cost Estimation.

Notation	Description
N	records number of whole dataset
C_{st}	start-up cost
C_{st}^m	mapper's start-up cost (optimization)
C_{st}^c	combiner's start-up cost (optimization)
C_w^m	cost spent on worker for executing mapper (optimization)
C_w^c	cost spent on worker for executing combiner (optimization)
C_{cl}	closure cost
C_w	cost spent on one worker
C_{mpg}	cost of creating a mapping from mapper to a worker node
C_{cmm}	communication cost
C_m	mapper's cost
C_r	reducer's cost
C_c	combiner's cost
C_s	one unit data's serialization cost
C_d	one unit data's de-serialization cost
C_l	cost of loading a record into memory
C_n	network factor, cost of transferring a unit of data
C_f	cost of filtering a record
C_a	cost of aggregating a record
C_i	total cost of initial implementation

Table 6.2: Notations Used for Describing the Cost Estimation (continued).

Notation	Description
$size_m$	size of mapper object
$size_c$	size of combiner object
$size_{rslt}$	size of mapper's intermediate result
nb_{GB}	Group-by dimension number
DV_i	the i_{th} distinct value
nb_m	mapper number
nb_{node}	worker node number
$\frac{N}{nb_m}$	block size
S	query's selectivity

the mapper is finished, it serializes the mapper's output. Therefore, the cost of this procedure is estimated as:

$$C_w = C_d \cdot size_m + C_m + C_s \cdot size_{rslt}$$

Similarly, let's assume that the de-serialization cost is proportional to $size_m$, and the cost of serializing the intermediate result generated by a mapper is proportional to $size_{rslt}$ by assumption.

The closure consists of de-serializing the mappers' outputs and executing the user-defined reducer. The de-serialization is run over all the records filtered by user-defined conditions in the query. Therefore, we know the total record number contained in all intermediate outputs to be serialized is equal to the number of all the filtered records: $N \times S$, (S represents the query's selectivity). Thus, we estimate the closure cost as below:

$$C_{cl} = C_d \cdot N \cdot S + C_r$$

The communication cost is composed of two parts. One is the cost of sending mappers from the master node to worker nodes; the other is that for worker nodes sending intermediate output to the master node. The size of messages and the network status are two factors considered in this cost estimation. Thus, we estimate the communication cost as:

$$C_{cmm} = C_n \cdot (nb_m \cdot size_m + N \cdot S)$$

The estimations of C_m and C_r are related to various applications. In the MapReduce-based initial implementation, the mappers perform filtering operations. They first load the data block from the disk into the memory, and then filter loaded data with condition defined in the query. For the mappers used in initial implementation, the cost estimation of the C_m is as follows:

$$C_m = \frac{N}{nb_m} \cdot (C_l + C_f \cdot S)$$

The reducer aggregates over the records filtered by mappers. It concerns the number of records that it processes. We estimate the reducer's cost as below:

$$C_r = N \cdot S \cdot C_a$$

With the mapper and reducer's cost estimation, we obtain the total cost estimation of the initial implementation. We consider that mapper object size is small compared with the intermediate outputs, and it can be ignored when the dataset is large; the costs concerning mappers' mapping, serialization, de-serializations and transmission can be removed. Thus, the estimation as below is obtained:

$$C_i = \frac{N}{nb_m} \cdot (C_l + C_f \cdot S) + C_s \cdot size_{rslt} + (C_d + C_a + C_n) \cdot N \cdot S \tag{1}$$

6.6.2 MapCombineReduce model-based implementation

For MapCombineReduce-based optimization of multiple-group-by query, the total cost is considered to be composed of:

- Mapper's start-up (on master), denoted as C_{st}^m;

- Cost spent on one worker executing a mapper, denoted as C_w^m;

- Combiners' start-up (on master), denoted as C_{st}^c;

- Cost spent on one worker executing a combiner, denoted as C_w^c;

- Closure (on master), denoted as C_{cl};

- Communication, denoted as C_{cmm}.

The start-up of mappers is similar to that of MapReduce; we mark two superscripts m and c in order to distinguish mapper's start-up from combiner's start-up:

$$C_{st}^m = (C_{mpg} + C_s \cdot size_m) \cdot nb_m$$

The mappers do the same calculations as in the initial implementation. However, the output size of each mapper is estimated as 0, because the mapper stores the selected records into the worker's memory and returns null. Thus, the cost for running a mapper is estimated as below:

$$C_w^m = C_d \cdot size_m + C_m + 0$$

where

$$C_m = \frac{N}{nb_m} \cdot (C_l + C_f \cdot S)$$

The combiner's start-up is similar to the mapper's start-up; however, the number of combiners is equal to the number of worker nodes, in that the combiners collect the intermediate data from all the worker nodes, so one combiner per worker node is sufficient. Thus, the combiner's start-up cost is estimated as below:

$$C_{st}^c = (C_{mpg} + C_s \cdot size_c) \cdot nb_{node}$$

The combiner's execution over one worker node can be estimated similarly as in the estimation of the mapper's execution. However, the size of the combiner's result can be precisely estimated as $\sum_{i=1}^{nb_{GB}} DV_i$, which is the result size of pre-aggregations on any worker node. Thus, we have the following estimation for combiner's execution cost:

$$C_w^c = C_d \cdot size_c + C_c + C_s \cdot \sum_{i=1}^{nb_{GB}} DV_i$$

where the combiner's cost is estimated as:

$$C_c = \frac{N}{nb_{node}} \cdot S \cdot C_a$$

The closure includes the de-serialization of the combiners' output and the cost of reducer:

$$C_{cl} = C_d \cdot \sum_{i=1}^{nb_{GB}} DV_i \cdot nb_{node} + C_r$$

where the reducer's cost is estimated as:

$$C_r = nb_{node} \cdot \sum_{i=1}^{nb_{GB}} DV_i \cdot C_a$$

As an additional combiner is added, the communication cost's estimation is correspondingly modified:

$$C_{cmm} = C_n \cdot (nb_m \cdot size_m + nb_{node} \cdot size_c + \sum_{i=1}^{nb_{GB}} DV_i * nb_{node})$$

The following estimation of total cost is obtained after ignoring the mapping, serialization/de-serialization and transmission cost of mappers and combiners:

$$C_o = \frac{N}{nb_m} \cdot (C_l + C_f \cdot S) + \frac{N}{nb_{node}} * S \cdot C_a +$$

$$(C_n + C_d + C_a) \cdot nb_{node} \cdot \sum_{i=1}^{nb_{GB}} DV_i + C_s \cdot \sum_{i=1}^{nb_{GB}} DV_i \quad (2)$$

6.6.3   Comparison of implementations

Comparing the equations (1) and (2), we can see that the optimized implementation surpasses the initial one in two aspects. First, it decreases the communication cost by reducing it from the scale of $N \times S$ to $nb_{node} \cdot \sum_{i=1}^{nb_{GB}} DV_i$. Second, a part of aggregating calculations is parallelized over worker nodes. We call the aggregation parallelized over worker nodes pre-aggregation. The aggregating phase's calculation in initial implementation had a scale of $N \times S$, and it is reduced to $(\frac{N}{nb_{node}} \cdot S + nb_{node} \cdot \sum_{i=1}^{nb_{GB}} DV_i)$ in the optimized implementation. However, another part of aggregation (post-aggregation) is inevitably to be done by master node; fortunately, the post-aggregation is small relative to the whole aggregation.

A disadvantage of the optimized implementation can be observed. A part of the cost is increased with the growth in worker node number, including communication cost. With this knowledge, the compression of intermediate output is considered to be important.

6.7   Concluding remarks

In this chapter, we first introduced the data explorer background of this work and identified the *Multiple Group-by* query as the elementary computation to be parallelized under this background. Then we described why we chose GridGain over Hadoop as the MapReduce framework in our work. We used GridGain as the MapReduce supporting framework because of its low latency. A detailed workflow analysis of the GridGain MapReduce procedure has been done. We realized two implementations of *Multiple Group-by* query based on MapReduce, initial and optimized implementations. The initial implementation of the *Multiple Group-by* query is based on a direct realization, which implemented the filtering phase within mappers and the aggregating phase within the reducer. In the optimized implementation of the *Multiple Group-by* query, we adopted a combiner as a pre-aggregator, which does the aggregation (pre-aggregation) on a local computing node level before starting the reducer. With such a pre-aggregator, the amount of intermediate data transferred over the network is reduced. As GridGain does not support a combiner component, we constructed the combiner through merging two successive GridGain's MapReduces. The experiments were run on a public academic platform named Grid'5000. The experimental results showed that the optimized version has better speed-up and better scalability for reasonable query selectivity. At the end of this chapter, a formal estimation of execution time is given for both implementations. A qualitative comparison between these implementations was presented. According to the qualitative comparison, the optimized implementation has decreased the communication cost by reducing the intermediate data; it has also reduced the aggregating phase's calculation by parallelizing a part of aggregating calculations. These estimations are also a valuable reference for other MapReduced applications.

Chapter 7

Multi-dimensional data analysis optimization

7.1 Introduction

In this chapter, we will present some methods to improve the performance of MapReduce-based *Multiple Group-by* query processing [Pan et al., 2010c, Pan et al., 2010a, Pan et al., 2010b]. In a distributed shared-nothing architecture, like the MapReduce system, there are two approaches to optimize query processing. The first one is to choose optimal job-scheduling policy in order to complete the calculation within minimum time. Load balancing, data skew, straggler node etc. are the issues involved in job-scheduling. The second approach focuses on the optimization of individual jobs constituting the parallel query processing. Individual job optimization needs to consider the characteristics of involved computations, including the low-level optimization of detailed operations. The optimization of individual jobs sometimes affects the job-scheduling policy. Although the two optimizing approaches are at different levels, they influence each other. In this chapter, we will first discuss the optimization work for accelerating individual jobs during the parallel processing procedure of the *Multiple Group-by* query. Then, we will identify the performance affecting factors during this procedure. The performance measurement work will be presented. The execution time estimation models are proposed for query executions based on different data partitioning methods. An alternative compressed data structure will be proposed at the end of this chapter. It enables one to realize more flexible job scheduling.

7.2 Data-locating based job-scheduling

GridGain is a Multiple-Map-One-Reduce framework. It provides an automatic job-scheduling scheme, which assumes all nodes are equally suitable for executing a job. Unfortunately, that is not the case in our work. We provide a data-locating job-scheduling scheme. This scheme can be simply described as sending a job to where its input data is.

7.2.1   Job-scheduling implementation

Our job-scheduling implementation helps the mapper to accurately locate data partition. This is especially important in case of no existence of data redundancy. One wrong mapping will cause computational errors. With the data placement procedure performed during data restructuring, this job-scheduling scheme is converted to a data location issue. We utilize the user-definable attribute mechanism provided by GridGain to address this issue. For example, we add a user-defined attribute "fragment" to each worker's GridGain configuration, and attribute it a value representing the data partitions' identifiers that it holds. When the worker nodes' GridGain instances are started, the "fragment" attribute is visible to the master node's GridGain instance and the other worker nodes' GridGain instances. It is used to identify the right worker node.

In the case of horizontal partitioning, worker node identifiers (i.e. hostnames) are utilized to locate data partitions. In this case, an equal number of partitions is placed on each worker node. The partitions containing successive records are placed over one worker node. That is, partitions are distributed on worker nodes in a sequential order. Then, the identifier of the worker node is used as the identifier of the data partitions that it holds. In this way, worker node identifiers are used to locate data partitions. For example, assuming that ROWSET is horizontally divided into ten partitions, these partitions are placed over five worker nodes. Thus, worker node A holds partitions one, two; worker node B holds partitions three, four, and so on. In this scenario, ten mappers need to be dispatched. As partitions one and two locate on worker node A, then mappers one and two are sent to worker node A. The rest of mappers are scheduled in the same way.

In case of vertical partitioning, a user-defined attribute, "region identifier" is utilized to locate data partitions. When a worker node number is small (case of one region), vertical partitions are replicated across all worker nodes. When a worker node number is large, further vertical partitions are horizontally divided into regions. Worker nodes are re-organized into regions accordingly. The worker nodes of the same region have the same "region identifier." Partitions are replicated across worker nodes within the same region. Thus, the region identifiers of worker nodes are utilized for data partitions. As an example, 13 vertical partitions from ROWSET, having 10,000,000 records, are horizontally divided into two regions. The records 1 to 5,000,000 are put in region one, and the records 5,000,001 to 10,000,000 are put in region two. Ten worker nodes are re-organized into two regions accordingly, each containing five worker nodes. Ten mappers aggregate over five different dimensions in two different regions respectively. For load balancing reasons, we use round-robin policy within regions to keep the job number running over each worker as balanced as possible.

7.2.2   Two-level scheduling

We actually realize two-level scheduling in MapReduce computations, i.e. task-level scheduling and job-level scheduling. Task-level scheduling means dispatching each mapper to the corresponding worker node. It considers how to distribute mappers, and ignores the calculation details within each job (mapper). Several elements should be considered in the task-level scheduling, such as mapper number, worker node number, and load balancing. In order to achieve load balancing, it is necessary to take into account worker node performance and status, and the input data location, etc. Because one job is run on one worker node, job-level scheduling takes place within a worker node, since one job runs on one worker node. Job-level scheduling considers the organization of calculations within a mapper. The main calculations can be encapsulated in reusable classes, and stored in a local jar file on each worker. The mapper calls the methods of these classes to run those calculations. Job-level scheduling is closely related to calculations that a job should execute. For this reason, the job-level scheduling should be tuned according to different queries; on the contrary, the task-level scheduling could be unchanged or slightly changed for different queries. Our mapper job definitions can be considered as a job-level scheduling.

7.2.3 Alternative job-scheduling schemes

An alternative job-scheduling scheme is to perform the filtering phase on the master node and the aggregating phase over worker nodes. This job-scheduling scheme is feasible since the restructured data allows loosely coupled computations. In the preceding implementation, the filtering phase and the aggregating phase are not separable, since aggregating phase computations consume filtering computations' output. With restructured data, we can see that the computations of these two phases are clearly decoupled, since they use different files as input data. In the filtering phase, a *search* operation is performed via accessing only inverted index files (Lucene generated files). In the aggregating phase, aggregation is performed over filtered records identified by a list of recordID calculated by filtering phase, and it only needs to access the compressed data files (FactIndex and Fact files). As these two phases are decoupled, they can be scheduled and optimized separately, which provides more flexibility for job-scheduling. This is especially helpful in case of vertical partitioning, where the selected recordIDs are commonly usable for multiple dimension aggregations.

7.3 Improvements by speed-up measurements

We evaluated our MapReduce-based *Multiple Group-by* query over restructured data in a cluster of Grid'5000 located in Orsay site[1]. Also, we used the version of GridGain 2.1.1 over Java 1.6.0. The JVM's maximum of heap size is set to 1536MB

---

[1]The cluster located in the Sophia site had unfortunately retired after doing our first part of experiments. The currently chosen cluster has the same hardware configuration as the retired Sophia cluster.

on both master node and worker nodes. We ran our applications over 1 to 15 nodes. Although the worker nodes were small-scaled, the ROWSET processed in these experiments is not extremely large, and it fits well with the amount of nodes used in our work. ROWSET was composed of 10,000,000 records with each including 15 columns. The size of ROWSET was 1.2 GB. We partitioned the dataset with both horizontal partitioning and vertical partitioning. All the partitions had already been indexed with Lucene and compressed before launching the experiments.

We chose queries having different selectivity. Selectivity is a factor that controls the amount of data being processed in the aggregating phase. Four *Multiple Group-by* queries' selectivities are 1.06%, 9.9%, 18.5% and 43.1% respectively. These queries all had the same five Group-by dimensions. Before starting the parallel experiments, we ran a group of sequential versions for each of these queries and measured the execution times, which were used as the baseline of the speed-up comparison.

7.3.1   Horizontal partitioning

Under the horizontal partitioning, we partitioned the ROWSET with different sizes. We ran, concurrently, different numbers of mappers over each worker node in different experiments. Thus, we could compare the performance of running a few big-grained jobs per node against that of running multiple small-grained jobs on one node. Our experiments with the horizontal partitioning-based implementation was organized in four groups. In the first group, there was only one mapper being dispatched to a worker node and run on it. In the second group, two mappers were running on one worker node. In the third group, we ran ten mappers on each worker node, and in the fourth group, twenty mappers per worker node. The Figure 7.1 shows the speed-up performance of the MapReduce-based *Multiple Group-by* query over horizontal partitions. We also realized a MapCombineReduce-based implementation. The MapCombineReduce-based implementation was for the case where more than one mapper was running on one node. The combiner performed the same computations as the reducer. The Figure 7.2 shows the speed-up performance measurement of the MapCombineReduce-based multiple Group-by aggregation over horizontal partitions.

Observation and comparison.   For the MapReduce-based implementation, the first observation of the speed-up measurement is that the queries with high selectivity shows better speed-up performance than the queries with low selectivity. A query with a certain selectivity has a fixed workload of calculation. Some parts of this workload are parallelizable, but others are not. The reason the high selectivity queries have better speed-up performance is that the parallelizable portion in their workload is greater than that in the low selectivity queries. The second observation is the speed-up performances of smaller job number per node (one and two jobs/node) experiments surpass that of bigger job number per node (ten and twenty jobs/node) experiments. Multiple jobs concurrently running over one node were considered to be able to utilize the CPU cycles more efficiently, and run faster. But in reality, this

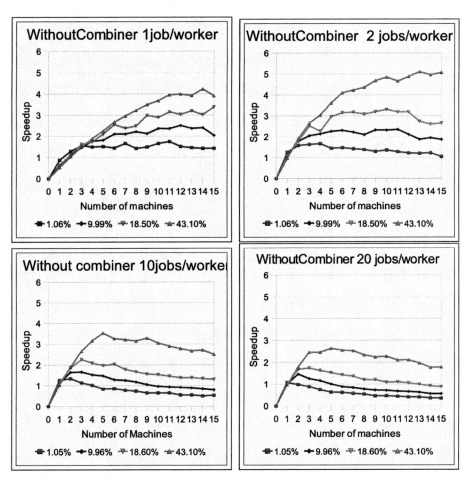

FIGURE 7.1: Speed-up of MapReduce *Multiple Group-by* query over horizontal partitions.

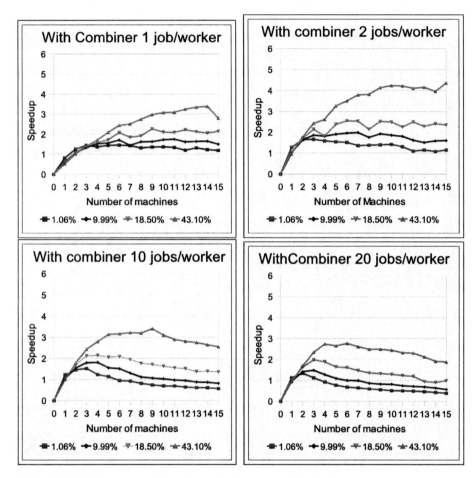

FIGURE 7.2: Speed-up of MapCombineReduce *Multiple Group-by* query over horizontal partitions.

is not always true. We will discuss the issue of multiple jobs concurrently running on one worker node later in this chapter.

The speed-up of MapCombineReduce-based implementation is similar to that of the MapReduce-based one. Comparing these two implementations, we can see that the speed-up performance of MapReduce-based implementation is better than that of the MapCombineReduce-based one in the experiments of small job number per node. In contrast, for experiments of big job number per node, the MapCombineReduce-based implementation speeds up better than the MapReduce-based one. That is due to the necessity of the combiner for different job number per node. For the job number per node, smaller or around the CPU number per node (e.g. one and two), the pre-final-aggregation (combiner's work) is not necessary, in that the number of intermediate outputs is not big. On the contrary, when the number of jobs per node is big (e.g. ten and twenty), the combiner is necessary. In this case, the speed-up of MapCombineReduce-based implementation is slightly better than the MapReduce-based implementation.

7.3.2   Vertical partitioning

Under vertical partitioning, we dispatched the vertical partitions using the policies described in Section 4.4.3. Similarly, we realized a MapReduce based implementation and a MapCombineReduce based one and measured the speed-up performance for both. During the experiments, we increased the number of worker nodes from 1 to 15, and divided the experiments into three groups. In experiments of group one, we had a small worker node number, denoted as w, $w \in [1..5]$; we organized vertical partitions into one region. If we note region number as nb_r, then $nb_r = 1$. In this case, each mapper aggregates over one entire Group-by dimension. Thus, in case of one region, the number of mappers is equal to the number of Group-by dimensions ($nb_m = nb_{GB} = 5$). In the second group of experiments, we increased the number of region to two ($nb_r = 2$) in order to utilize up to ten worker nodes. We ran the queries over 2, 4, 6, 8 then 10 worker nodes (i.e. $w \in [2, 4, 6, 8, 10]$), and measured the execution time in the case of each vertical partition being cut into two regions. As the number of the mappers equals the number of partitions, then we have $nb_m = nb_{GB} \cdot nb_r = 10$. In the third group of experiments, we increased the number of regions to three, i.e. $nb_r = 3$. We had worker nodes number $w \in [3, 6, 9, 12, 15]$ in different experiments. The number of mapper $nb_m = 15$. The mappers were evenly distributed within each region.

As we fixed the Group-by dimension number at five, the total mapper number was $5 \times nb_r$, and the number of mappers per node was varied with the node number per region: $nb_{job/node} = \lfloor 5/nb_{node/region} \rfloor$ or $\lceil 5/nb_{node/region} \rceil$. For example, if $nb_{node} = 1$, $nb_r = 1$, then each node was assigned five mappers; if $nb_{node} = 10$, $nb_r = 2$, then each node was assigned one mapper; if $nb_{node} = 2$, $nb_r = 1$, then one node was assigned two mappers, the other three mappers, etc. We illustrate the speed-up performance measurements in the Figure 7.3.

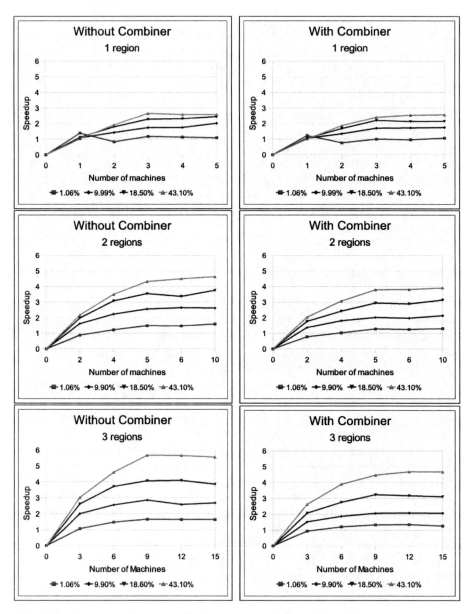

FIGURE 7.3:   Speed-up of MapReduce *Multiple Group-by* aggregation over vertical partitions.   MapReduce-based implementation is on the left side. MapCombineReduce-based implementation is on the right side.

Observation and comparison. As shown in Figure 7.3, the speed-up is increasing with the rise in worker number regardless of the number of regions. The MapReduce-based implementation speeds up better than the MapCombineReduce-based one, because the number of jobs per node is small (i.e. < 5). The queries with bigger selectivity, like 9.9%, 18.5%, 43.1%, benefit more from the parallelization than the queries with smaller selectivity, like 1.06%.

For most queries, the biggest speed-up appears in the third group of experiments with three regions, for both the MapReduce-based and MapCombineReduce-based implementations. Comparing the speed-up under vertical partitioning and that under the horizontal partitioning, we can see the best speed-up appears in experiments under the vertical partitioning. Under vertical partitioning, each mapper aggregates over one dimension only; the obtained intermediate output is the aggregates of one dimension. The size of the intermediate outputs using vertical partitioning is much smaller than those using the horizontal partitioning. Imagining a scenario where 10 worker nodes are available, under horizontal partitioning, one mapper works on one horizontal partition on one worker node. As each mapper aggregates over 5 dimensions, then, the number of intermediate aggregate tables from all the mappers are $10 \times 5 = 50$. Under vertical partitioning, 10 available workers are organized into 2 regions. Also, there are in total 10 mappers. But each mapper aggregates over one dimension. Thus the number of intermediate aggregate tables is exactly the same as the number of mappers, 10. If we simply suppose that an aggregate table of an arbitrary Group-by dimension is of size 20K, then 1,000K intermediate output is generated under horizontal partitioning, while 200K intermediate output is generated under vertical partitioning. Thus, with vertical partitioning, the intermediate data volume to be transferred is reduced with regard to the experiments using the horizontal partitioning.

7.4 Improvements by affecting factors

In this section, we will discover the performance affecting factors in the *Multiple Group-by query* processing. Some of them concern the computations themselves, others are related to exterior conditions, such as hardware, network, etc. Discovering these factors is helpful in locating the bottlenecks, and in turn increasing the system efficiency. The performance affecting factors addressed in this section include query selectivity, running multiple jobs over one worker node, hitting data distribution, intermediate output size, serialization algorithms, network status, combiner utilization as well as data partitioning methods.

7.4.1   Query selectivity

Query selectivity is a factor that controls the records filtered out during the selecting phase. Also, it determines the amount of data that the aggregating phase should process. For big selectivity, query aggregating phase takes up a majority of the whole calculation. In addition, the aggregating calculation is parallelizable. Thus the query with big selectivity benefits more from the parallelization than query with small selectivity. Query selectivity is sometimes related to workload skew. In some particular cases a query selects a lot of records from some partitions, but very limited records from the other partitions. Therefore, most aggregate operations are performed only on a part of worker nodes, while the other nodes remain idle, which causes the workload skew. This happens more frequently with range data partitioning than in other cases.

7.4.2   Side effects

In our experiments, we ran a different number of mappers on each worker node so as to measure different effects for acceleration. Intuitively, the more mappers concurrently running on one worker node, the more efficiently the CPU(s) should be utilized. However, the mappers' run is degraded when contentions are provoked. More importantly, this retards the execution of the reducer, since the reducer does not start until all mappers have been finished. Thus, from the point of view of the whole query processing, running multiple mappers on one worker node may degrade the performance of individual mappers. This will in turn degrade the whole MapReduce procedure. Table 7.1 shows a list of average execution times of one individual mapper when multiple mappers run concurrently on one node. The workload of each mapper was as follows: searching in the inverted index to filter the data partition and obtaining a list of recordIDs with the records associated satisfying the WHERE condition; aggregating over one vertical partition. The total record number of the partition is 3 333 333, and the number of records selected out accounts for 1% of the total records. These mappers were executed on one same worker node.

Table 7.1: Average Execution Time of Multiple Mappers' Jobs on One Node.

Job number on one node	One Mapper's Average Execution time (ms)
1	170
2	208
3	354
4	417
5	537

These mappers are concurrently running as different threads. They do not communicate among each other, and they have different inputs and outputs. This means that each mapper will bring new input data into the memory and generate its output data. The workload of data aggregation is typically data-intensive, and contentions may occur in different resources, such as the contentions of disk I/O or memory bandwidth. As shown in this table, when running only one mapper over one worker node, the execution time is relatively small (170 ms). When concurrently running 2 mappers over one worker node, the average execution time of one individual mapper is slightly drawn out (37 ms longer). When concurrently running 3 or more mappers over one worker node, the execution time shows a relatively long delay (from 184 ms to 367 ms). We can see that, on one worker node with 2 CPUs, having 2 concurrently running mappers, the average execution time is the most interesting. After that, when we continuously increased the number of mappers on the worker node, the more mappers concurrently running on one worker node, the longer the time an individual mapper takes.

7.5   Improvement by cost estimation

In this section, we give a cost estimation model for the execution time during the whole MapReduce-based query processing on the restructured data. For the sake of time limitation, we worked only on the cost estimation for the MapReduced-based implementation, and the cost estimation for MapCombineReduce-based implementation is not addressed in this work. The above-mentioned performance affecting factors and observation based on our experiments are maximally considered for constructing the cost estimation model.

We still consider the four factors of cost in a MapReduce procedure, start-up, mapper execution, closure and communication. In start-up, the master prepares the mappers, including mapping mappers to available worker nodes, then serializing mapper objects. The serialization for the first mapper object takes longer than the serializations for the other mapper objects. The formal cost estimation for start-up is as follows:

$$C_{st} = C_{mpg} \cdot nb_m + C_s \cdot size_m + C'_s \cdot (nb_m - 1) \cdot size_m$$

If there is no additional specification, the notations used in the formulas of this chapter can be referred to in Table 6.1. In the above formula, we estimate the cost of computing the mappings as $C_{mpg} \cdot nb_m$, the serialization time for the first mapper object as $C_s \cdot size_m$, the serialization time for the rest of mapper objects as $C_s \cdot (nb_m - 1) \cdot size_m$

When a worker receives a message from the mapper, it de-serializes the mapper object, then executes mapper job. When finished, it serializes the aggregate table produced by the mapper. Taking into account the factor of running multiple mappers on one worker node, we add a function of mapper number per node (denoted as $f(nb_{m/node})$) into the estimate. Thus, the execution time of this process is estimated as:

$$C_w = f(nb_{m/node}) \cdot (C_d \cdot \gamma \cdot size_m + C_m + C_s \cdot size_{agg})$$

Here, $\gamma \cdot size_m$, $(\gamma > 1)$ is used to represent the size of the serialized mapper object. A serialized object is always bigger than the original one, so, we have $\gamma > 1$. Also, γ varies according to the composition of the object. The notation $size_{agg}$ means the size of the generated aggregate table. The mapper execution cost, C_m, and aggregate table size, $size_{agg}$, vary according to the adopted partitioning methods. We will respectively give the detailed estimations for the horizontal partitioning-based implementation and the vertical partitioning-based one at a later stage.

The closure includes de-serialization of the intermediate aggregate tables and user-defined reducer execution. We estimate the closure cost as below:

$$C_{cl} = C_d \cdot \sum_{i=1}^{nb_m} \gamma \cdot size_{agg\_i} + C_r$$

where the reducer cost (denoted as C_r) varies with different applications. We will give the estimation of reducer later.

In our work, the data communication is composed of the master node sending mappers to the worker nodes, and worker nodes sending intermediate aggregate tables to the master node. Considering the size of transmitted data and the network status we estimate the communication cost as:

$$C_{cmm} = C_n \cdot (nb_m \cdot \gamma \cdot size_m + \sum_{i=1}^{nb_m} \gamma \cdot size_{agg\_i})$$

where $size_{agg\_i}$ represents the size of the ith aggregate table produced by mappers.

7.5.1   Horizontal partitioning

For the implementation over horizontal partitions, the mapper takes a horizontal partition as input data, searches in its Lucene index, reads values of dimensions and aggregates with measures over the distinct values of Group-by dimensions. We assume there are D dimensions and M measures in ROWSET, over which we run the query aggregating on nb_{GB} Group-by dimensions. The mapper cost is estimated as below:

$$C_m = S \cdot \frac{N}{nb_m} \cdot \{\alpha \cdot C_f + \beta \cdot (nb_{GB} + 4M) \cdot C_{rd} + nb_{agg} \cdot nb_{GB} \cdot C_a\}$$

where C_f estimates the average execution time for successfully obtaining one recordID of the selected records by searching Lucene index; C_{rd} represents the execution time to retrieve 1 byte from the compressed file; nb_{agg} means the number of aggregate functions defined in the query. A distinct value is represented as an integer (i.e.

distinct value code) of size 1 byte[2], and a measure value as a float sized 4 bytes. As mentioned before, record filtering and record reading operations are impacted by hitting data distribution issues, which means the average time for processing one unit of data is varying with query selectivity. Therefore, two parameters α and β are applied over the corresponding items. Their values vary with different queries.

Under horizontal partitioning, each mapper produces aggregate tables for all Group-by dimensions. The size of the aggregate table can be estimated as follows:

$$size_{agg} = \sum_{i=1}^{nb_{GB}} nb_{DV_i} \cdot (1 + 4nb_{agg})$$

where nb_{DV_i} represents the number of distinct values of the ith Group-by dimension. $1 + 4nb_{agg}$ is the number of bytes contained in one row of the aggregate table.

As the reducer takes all intermediate outputs of mappers as input and performs aggregation over them, we estimate cost of the reducer as:

$$C_r = C_a \cdot nb_{agg} \cdot nb_m \cdot \sum_{i=1}^{nb_{GB}} nb_{DV_i} \tag{1}$$

With this detailed estimate, we ignore the function $f(nb_{m/node})$, since we address the small job per node cases, which are the most common cases. Thus, we obtain the total execution time estimate of the horizontal partitioning-based *Multiple Group-by* query as below:

$$Cost_{hp} = C_{mpg} \cdot nb_m + C_s \cdot size_m + C'_s \cdot (nb_m - 1) \cdot size_m + C_d \cdot \gamma \cdot size_m +$$

$$S \cdot \frac{N}{nb_m} \cdot [\alpha \cdot C_f + \beta \cdot (nb_{GB} + 4M) \cdot C_{rd} + nb_{agg} \cdot nb_{GB} \cdot C_a] +$$

$$C_s \cdot \sum_{i=1}^{nb_{GB}} nb_{DV_i} \cdot (1 + 4nb_{agg}) + C_d \cdot nb_m \cdot \sum_{i=1}^{nb_{GB}} nb_{DV_i} \cdot (1 + 4nb_{agg})$$

$$+ C_a \cdot nb_{agg} \cdot nb_m \sum_{i=1}^{nb_{GB}} nb_{DV_i} + C_n \cdot nb_m \cdot \gamma \cdot size_m + C_n \cdot \gamma \cdot \sum_{i=1}^{nb_m} size_{Sagg_i}$$

If we note the average size of the serialized aggregate table as $\gamma \cdot \sum_{i=1}^{nb_{GB}} nb_{DV_i} \cdot 1 + 4nb_{agg}$. We estimate the value of parameters as described in table 7.2. As it is difficult to accurately estimate the serialization/de-serialization execution time for a unit of data, we estimate the serialization/de-serialization time for the data really used in our experiments. Therefore, the estimated values for these parameters are accompanied by the size of data being serialized or de-serialized.

In order to test the accuracy of our execution time estimation model, we compare the speed-up curve calculated from our model to the measured speed-up curve.

---

[2]In our work, the distinct value number of any dimension is smaller than 256, thus, 1 byte is sufficient to represent all the distinct value codes in integers of any dimension.

Table 7.2: Parameters and Their Estimated Values (in ms).

Notation	Estimated value	Cost for...
C_{mpg}	2.34×10^{-1}	creating a mapping between mapper and a worker node
$C_s \cdot size_m$	83.51	serializing first mapper instance
$C_s' \cdot size_m$	1.21	serializing non-first mapper instance
$C_d \cdot \gamma \cdot size_m$	2.45	de-serializing mapper
C_a in mapper	0	aggregation, ignorable, since we use small number aggregate functions (only 2) in our work; aggregate operation is right after retrieving the operand.
C_a in reducer	0.001	aggregating in reducer
C_s	6.67×10^{-3}	serializing one byte of aggregate table on average in mapper
$C_d \cdot \gamma$	5.0×10^{-3}	de-serializing for one byte of aggregate tables on average in reducer
$C_n \cdot \gamma \cdot size_m$	0.403	transmitting one mapper
$C_n \cdot \gamma$	8.82×10^{-4}	transmitting one byte of aggregate table
For query $selectivity = 1.06\%$		
$\alpha \cdot C_f$	7.24×10^{-4}	filtering per record on average
$\beta \cdot C_r d$	4.33×10^{-4}	reading one byte from compressed data
For query $selectivity = 9.9\%$		
$\alpha \cdot C_f$	1.15×10^{-4}	filtering per record on average
$\beta \cdot C_r d$	7.20×10^{-5}	reading one byte from compressed data
For query $selectivity = 18.5\%$		
$\alpha \cdot C_f$	5.30×10^{-5}	filtering per record on average
$\beta \cdot C_r d$	7.60×10^{-5}	reading one byte from compressed data
For query $selectivity = 43.1\%$		
$\alpha \cdot C_f$	5.30×10^{-5}	filtering per record on average
$\beta \cdot C_r d$	7.60×10^{-5}	reading one byte from compressed data

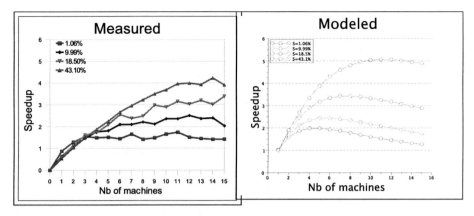

FIGURE 7.4: Measured speedup curve versus a modeled speedup curve for MapReduce-based query on horizontal partitioned data, where each work node runs one mapper.

Figure 7.4 shows two speed-up curves for MapReduce-based query processing on horizontal partitions, with the number of worker nodes gradually increasing from 1 to 15. We chose the case with only one mapper concurrently running on a worker node. In this case, the application related parameter can be determined, such as, total record number N=10,000,000, aggregate function number $nb_{agg} = 2$, total distinct values number $\sum_{i=1}^{nb_{GB}} nb_{DV_i} = 511$, Group-by dimension number $nb_{GB} = 5$, measure values contained in one record $M = 2$, etc.

7.5.2 Vertical partitioning

We also estimate the cost of the implementation under the vertical partitioning in a similar way. The mapper cost is estimated as:

$$C_m = S \cdot \frac{N}{nb_{rgn}} \cdot [C_f + (4M + 1) \cdot C_{rd} + C_a \cdot nb_{agg}]$$

where nb_{rgn} means the number of regions.

With vertical partitioning, each mapper aggregates over one dimension d then the intermediate aggregate table size is estimated as:

$$size_{agg_d} = nb_{DV_d} \cdot (1 + 4nb_{agg})$$

where, $nb_{nb_{DV_d}}$ means the distinct value number of current Group-by dimension d; $(1 + 4nb_{agg})$ is the estimated size in byte of each row in aggregate table.

The reducer aggregates over a list of aggregate results; each of them is the aggregate result of one dimension or part of dimension in the case of $nb_{rgn} > 1$. The estimation of the reducer is:

$$C_r = C_a \cdot nb_{rgn} \cdot nb_{agg} \cdot \gamma \cdot \sum_{i=1}^{nb_{GB}} nb_{DV_i} \qquad (2)$$

By summing up the above estimations, we obtain the total cost estimate of *Multiple Group-by* query processing over vertical partitions:

$$Cost_{vp} = C_{mpg} \cdot nb_m + C_s \cdot size_m + C_d \cdot \gamma \cdot size_m +$$

$$S \cdot \frac{N}{nb_{rgn}} \cdot [C_f + (4M + 1) \cdot C_{rd} + nb_{agg} \cdot C_a] +$$

$$C_s \cdot nb_{DV_d} \cdot (1 + 4nb_{agg}) + C_d \cdot \gamma \cdot nb_{rgn} \cdot \sum_{i=1}^{nb_{GB}} nb_{DV_i} \cdot (1 + 4nb_{agg}) +$$

$$C_a \cdot nb_{rgn} \cdot nb_{agg} \sum_{i=1}^{nb_{GB}} nb_{DV_i} + C_n \cdot (nb_m \cdot \gamma \cdot size_m + \gamma \cdot nb_{rgn} \cdot \sum_{i=1}^{nb_{GB}} nb_{DV_i} \cdot (1 + 4nb_{agg}))$$

Here we replaced $\sum_{i=1}^{nb_m} size_{agg_i}$ by $nb_{rgn} \cdot \sum_{i=1}^{nb_{GB}} nb_{DV_i} \cdot (1 + 4nb_{agg})$ since the partial aggregate tables from different regions effectively construct nb_{rgn} times aggregates for all Group-by dimensions. The same estimation for parameter values could also be done for vertical partitioning base query processing. We will have this done in the future work.

7.5.3   Comparison of partitioning

Note that, for the same ROWSET partitioned horizontally and vertically, the number of partitions in horizontal partitioning is larger than the number of regions in vertical partitioning, that is, $HP.nb_{pttn} > VP.nb_{rgn}$. The reason is that in horizontal partitioning, the partition number is equal to mapper number, and we let the mapper number equal a multiple of the node number, so as to utilize all the available nodes. However, with vertical partitioning, the region number is usually a sub-multiple of the nodes number. Given this established fact, $HP.nb_{pttn} > VP.nb_{rgn}$, we see that the reducer cost of vertical partitioning-based implementation, which is expressed in formula (1), is smaller than that of the horizontal partitioning-based one, which is expressed with formula (2). As shown by the estimation, under both horizontal partitioning and vertical partitioning, a great part of calculation is parallelized. We do the calculation, reduced from scale of ROWSET size N to fragment size $\frac{N}{nb_m}$ (in case of horizontal partitioning) or $\frac{N}{nb_{rgn}}$ (in case of vertical and hybrid partitioning). However, the transfer and serialization/de-serialization of intermediate data still forms an important part of the cost. This cost results from parallelization. On the contrary, we can imagine that a further compression of mappers for intermediate outputs could optimize the calculation.

7.6   Compressed data structures

In the previous work, we used a data partition locating policy as the job-scheduling policy. Although this worked well, we still need a more flexible job-scheduling policy. An imaginable job-scheduling policy is based on distinct values. That means, each mapper works to aggregate only one or a part of distinct values of one certain dimension, then the intermediate aggregate tables produced by the mappers are assembled in the reducer. To support such a distinct-value-wise job scheduling, we propose an alternative compressed data structure in this section. This data structure works with vertical partitioning.

7.6.1   Data structure description

In order to facilitate distinct-value-wise job scheduling, we need to be able to calculate the aggregate value of one distinct value within one continuous process. Thus, if the measure values corresponding to the same distinct value are successively stored, then aggregation for one distinct value can be processed in a continuous mode. This is the basic idea of the new compressed structure. In this new compressed data structure, we regroup the measure values' storage order. Measures corresponding to the same distinct value are stored together successively. As the stored order of measures is different than in the original ROWSET, we provide a data structure recording the records' former positions in the original ROWSET. Relying on the above description, we design the compressed data structure as follows. To be noted, this structure is designed specifically for vertical partitioning. The compressed data is still composed of two files, *FactIndex* and *Fact*. For each distinct value, Fact file stores a recordID-list with each recordID indicating the former position of records containing the current distinct value. It then stores a set of measures containing the current distinct value. FactIndex stores for each distinct value, the distinct value code, and an address pointing to a position in the Fact file where the recordID-list and a set of measures covered by the current distinct value start to store. Figure 7.5 illustrates this structure.

For aggregating using this data structure, each mapper will be scheduled to aggregate over one distinct value. FactIndex file is accessed to obtain the given distinct value's storage position in the Fact file. Then, the mapper identifies the selected records covered by the given distinct values by retrieving the common recordIDs of selected recordID-list from the filtering phase and the recorID-list retrieved from the Fact file. Finally, the selected records covered by the current value are aggregated using the measure values covered by current distinct value retrieved from the Fact file.

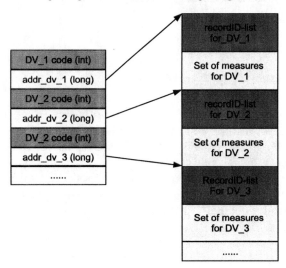

FIGURE 7.5: Compressed data files suitable for distinct value level job scheduling with measures for each distinct value stored together.

7.6.2   Data structures for storing recordID-list

Integer Array and Bitmap are two alternative data types which can be used to store a recordID-list. In the case of using Integer Array, each recordID is stored as an element of the array. In the case of using Bitmap, we create a Bitmap composed of a sequence of bits. The number of bits is equal to the cardinality of the original ROWSET. One bit in Bitmap corresponds to one record. The value of each bit is either 0 or 1. If we use 1 to indicate that the current record I.D. is in the current distinct value's recordID-list, we obtain a Bitmap with all "1" positions indicating the whole recordID-list.

Regarding the use of storage space, Integer Array and Bitmap are very different. When the recordID-list contains a small number of elements, Integer Array takes smaller storage spaces. In the opposite case, where the recordID-list contains a large number of elements, Bitmap is more storage efficient.

7.6.3   Compressed data structures for different dimensions

Taking into account the above features, we distinguish two categories of dimensions: dimensions having a small number of distinct values and dimensions having a large number of distinct values. For those dimensions with a small number of distinct values, many records are covered by one certain distinct value. In turn, a large number of recordIDs need to be stored. In this case, Bitmap is more space-saving and makes access more efficient than Integer Array. For those dimensions with a large number of distinct values, only a few recordIDs need to be stored. For this

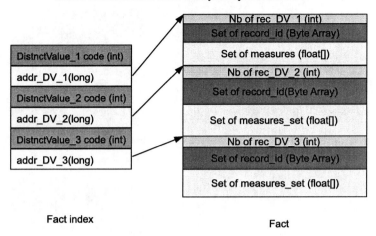

Fact index   Fact

FIGURE 7.6: Compressed data structure storing recordID list as an integer array for dimensions with a large number of distinct values.

case, Integer Array is more space-saving and provides more access efficiency.

But how to define the "small" and "large" over the number of distinct values for one dimension? Let's do a concrete calculation. Imagine that we have a ROWSET containing 10^7 records. One recordID stored as an integer takes 4 bytes. Assume that a dimension includes V distinct values. Thus each distinct value covers $10^7/V$ records. If we store the recordID-list in Integer Array, for one certain distinct value, $4 \times 10^7/V$ bytes are needed, on average. Then, we need to store $4 \times 10^7/V$ bytes in total in order to store all recordIDs containing all distinct values. If we use a Bitmap to store a recordID-list of one distinct value, then the Bitmap takes 1.25×10^6 bytes. As a result, the critical point of distinct value number V is 32. If $V = 32$, then two units of storage take up the same space; if $V < 32$, Bitmap takes up a smaller space; if $DV > 32$, then Integer Array takes up a smaller space. Thus, if the number of distinct values is larger than 32, then we consider it as "large"; otherwise, if the number of distinct value is smaller than or equal to 32, then we consider it as "small." After defining the data structure of a recordID-list, we specify the concrete storage for those two dimensions. For dimensions having a large number of distinct values, the composed data is composed of 2 files, FactIndex and Fact. Data stored in the FactIndex file includes the code of each distinct value of an integer and an address of a long integer pointing to a position in the Fact file, where the data related to this distinct value is stored. Data stored in the Fact file includes three parts. The first one is an integer representing the number of records covered by the current distinct value. The second one is an Integer Array compressed in Byte Array representing the recordID-list for records covered by the current distinct value. The third one is a Float Array representing the measure values for records having the current distinct value. Refer to Figure 7.6 for the illustration of this structure.

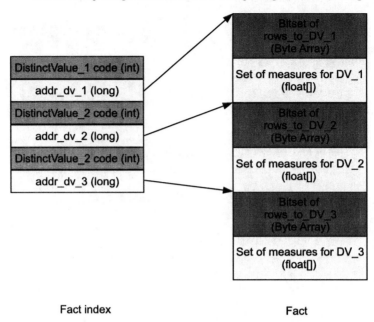

Fact index   Fact

FIGURE 7.7: Compressed data structure storing recordID list as Bitmap for dimension with a small number of distinct values (In Java, Bitmap is implemented as Bitset)

For dimensions having a small number of distinct values, the compressed data is also composed of a FactIndex file and a Fact file. The FactIndex file stores the code of the current distinct value in an integer and an address in a long integer, in long type, pointing to a position in the Fact file where data related to this distinct value is stored. The Fact file stores, for each distinct value, a Bitmap indicating the records covered by the current distinct value, as a Byte Array, and the measure values for records having the current distinct value in a type of Float Array. Refer to Figure 7.7 for the illustration of this structure.

7.6.4   Bitmap sparcity and compressing

The low efficiency of Bitmap, i.e. the second compressed data used in the above experiments is caused by the sparcity. Even for a dimension having a small number of distinct values, for example, 12, the Bitmap is very sparse, since only $1/12$ bits are set to 1. A Bitmap compressing is crucial to improving the storage efficiency. There are already some Bitmap compressing algorithms that we can take advantage of. These methods typically employ Run-length-encoding, such as Byte-aligned Bitmap Code, Word-Aligned Hybrid code and Position List Word Aligned Hybrid [Bitmap, 2012]. Run-length-encoding stores the sequence in which the same data value occurs in many consecutive positions, namely run, as one single data value and count, instead of storing them as the original run [Run-length-encoding, 2012]. We will address the Bitmap compressing methods in future work in order to improve our Bitmap's storage efficiency.

7.7   Concluding remarks

In this work, we realized *Multiple Group-by* query on restructured data, using the MapReduce model to parallelize the calculation. We introduced data partitioning, indexation, and data compression processing in the data restructuring phase. The materialized view ROWSET is partitioned using two principal partitioning methods, horizontal partitioning and vertical partitioning, respectively. The index that we created using Lucene over ROWSET is an inverted index, which allows rapid accessing and filtering of the records with the WHERE condition. We measured our *Multiple Group-by* query implementations over ROWSET, and compared the speed-up performance of implementations over horizontally partitioned data and that of vertically partitioned data. In most cases, they showed similar speed-up performance; however, the best speed-up appeared in the vertical partitioning-base implementation. Based on the measured result observations and analysis, we discovered several interesting factors that affect query processing performance, including query selectivity, concurrently running mapper numbers on one node, hitting data distribution, intermediate output size, adopted serialization algorithms, network status, whether or not using combiner as well as the data partitioning methods. We gave an estimation model for the query processing execution time, and specifically estimated the values of various parameters for data horizontal partitioning-based query processing. In order to support distinct-value-wise job-scheduling, we designed a new compressed data structure, which works with vertical partition. It allows the aggregations over one certain distinct value to be performed within one continuous process. However, such a data structure is only an initial design. We will address the optimization issues, like Bitmap compressing, in future work.

Chapter 8

Real-time scheduling with MapReduce

8.1 Introduction

MapReduce [Dean and Ghemawat, 2008] has emerged as one of the most popular frameworks for distributed cloud computing. The simple but powerful programming model is beneficial to a wide spectrum of data-intensive applications such as search indexing, mining social networks, recommendation services, and advertising backends. These applications enable computing datacenter support in carrying out daily activities as well as solving social problems. Since computing is becoming more instrumented, interconnected, intelligent and pervasive than ever before, it brings many challenges in systems design, modeling and engineering. There are emerging classes of cloud-based applications that can benefit from increasing time guarantee. For example, real-time advertising requires a real-time prediction about user intent based on their search histories. Meeting deadlines here can translate into higher profits for the content providers. In control datacenters, enormous amounts of real-time data need to be collected and reported periodically by various sensors. Besides that, ambient intelligence needs a networked database to integrate these sensor data streams in time and to give a real-time analysis result according to event requests. Therefore, computing in clouds, where billions of events occur simultaneously, is not in a time linear dimension, but falls into the real-time computing category.

Real-time application is subject to a real-time constraint that must be met, regardless of system load. If a real-time computation does not complete before its deadline, it is treated as a failed case, as serious as if the computation were never executed. Dealing with different real-time tasks on a MapReduce cluster can benefit users sharing a common large dataset. However, the traditional scheduling schemes need to be revised in terms of particular characteristics of MapReduce.

8.2   Real-time scheduling problem

We build a real-time scheduling problem model by a triple (Γ, P, A) where Γ is the set of real-time tasks, P the set of processing resources and A the scheduling algorithms.

8.2.1   Real-time task

A computing task is an application taking up memory space and execution time. The concept of task should be distinguished from event. An event emphasizes an operation taking place at a specific moment, while a task can be submitted, executed, halted, suspended and returned.

For the purpose of time analysis, we define a real-time task by its timing character-istics, rather than by the functionality requirements, such as execution time, lateness, deadline, etc. The tasks to be scheduled make a task set $\Gamma = \{\tau_1, \tau_2, \cdots, \tau_n\}$, and any τ_i consists of a periodic sequence of requests. When a request arrives, a new instance is created. For the periodic real-tasks, several preliminary terms should be defined.

- T_i: period is the time between two successive instances of task τ_i.

- O_i: offset is the first release of task τ_i.

- C_i: computation time is the worst-case execution time of τ_i.

- D_i: deadline is the relative overdue time in one period.

In addition, $\tau_{i,k}$ denotes the kth instance of τ_i. There are several important instances for $\tau_{i,k}$, and their relationship as shown in Figure 8.1

- $a_{i,k}$: activation instant at which instance $\tau_{i,k}$ is released to the scheduler.

- $s_{i,k}$: starting instant at which the instance $\tau_{i,k}$ starts computation.

- $e_{i,k}$: execution time; it is how long instance $\tau_{i,k}$ is running

- $f_{i,k}$: finishing instant at which instance $\tau_{i,k}$ finishes the execution.

- $d_{i,k}$: overdue instant at which instance $\tau_{i,k}$ is required to be finished.

All instances are activated after the request is submitted, so $a_{i,k}$ is equal to $O_i + (k-1)T_i$. The starting time $s_{i,k}$ can not be earlier than the activation $a_{i,k}$. The total amount of execution time $e_{i,k}$ depends on the processing resources, but it can not exceed the worst execution time, that is $C_i = \max e_{i,k}$. The execution of $\tau_{i,k}$ finishes at $f_{i,k}$, and $s_{i,k} + e_{i,k} \leq f_{i,k}$. For most cases, the equal sign is not true, because the scheduler might execute more than one task at the same time. Finishing time is important, but varies with different instances. Response time of task τ_i is

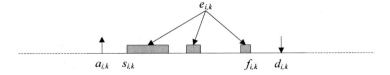

FIGURE 8.1: Relationship between important instants.

defined as the maximum of finishing time $R_i = \max(f_{i,k} - a_{i,k})$. The deadline $d_{i,k}$ is the absolute overdue time for $\tau_{i,k}$, so $d_{i,k} = a_{i,k} + D_i$. The task utilization $u_i = C_i/T_i$ shows the impact of task τ_i on processing resource. System utilization is the sum of all u_i, and it presents the fraction of processor time used by a periodic task set.

$$U = \sum_{i=1}^{n} u_i \tag{8.1}$$

Since the required amount of computation power can not exceed the available resource, the condition $U \leq 1$ must be satisfied if there are feasible scheduling solutions on task set Γ.

8.2.2 Processing resource

The processing resource is the resource in charge of executing tasks. For the sake of simplicity, we distinguish processors one from another by their computing capability. The concrete processor types or internal architectures are ignored in this model. Typical processing resources are

- Uniprocessor: there is only one processor in the set, and the worst-case computation time depends on the size of executed tasks.

- Identical multiprocessor: the number of processors in the set is more than one, and each of them has the same computing capability.

- Uniform multiprocessor: the number of processors in a set is more than one. Different processors have different computing capability, but the speed of each processor is a constant and does not depend on task type.

- Heterogeneous multiprocessor: multiprocessors are made up of different hardware platforms, so the worst-case computation time depends not only on task size, but also on task type.

Among them, the uniprocessor and the identical multiprocessor are most studied, because they are more general and easily analyzed than the multiprocessor of identical or heterogeneous configuration. The other two cases can be extended by identical multiprocessors. In particular, many results achieved for the uniprocessor are useful for multiprocessor resources; we therefore focus the discussion on the uniprocessor.

8.2.3   Scheduling algorithms

Scheduling algorithm A is the set of rules for mapping tasks from Γ onto the processing resource P. An algorithm is preemptive if the execution of one task can be interrupted by another task. The interrupted one is resumed later at the same location where the task was preempted. Non-preemptive algorithms are easily implemented because no extra overhead is needed for a context switch, but they can not promise that all deadlines are satisfied. As a result, preemptive algorithms are applied by real-time scheduling to handle applications with strict time requirements.

Two basic constraints should be met. A task can not be executed on two or more processors simultaneously, and a processor can not execute on two or more tasks. Under these premises, a feasible scheduling algorithm is that the scheduling can make all tasks meet their deadlines. An algorithm is optimal in the sense that no other feasible scheduling exists if the task set can not be scheduled by this algorithm.

The First In First Out (FIFO) algorithm queues tasks on a waiting list. When a new task is submitted, scheduler puts it on the list according to its arrival time. Round-Robin (RR) is another common scheduling algorithm. It handles all tasks without priority, and circularly assigns a fixed time unit to each task in equal portions. However, both of them perform badly in a real-time scheduling systems, which means they often fail to match the applications' constraints.

In the context of real-time systems, the scheduling algorithm is priority driven. The tasks are assigned priorities according to their constraints, and generally the highest priority is assigned to the most urgent task. When a task with low priority encounters another task with high priority, the running one immediately hands over the processor to the new task. Thus, the task with the highest priority is always executed whether the processor is occupied or not, using preemption if necessary. In this case, a static scheduling algorithm refers to fixed priority assignment. Once the priority is fixed, it never changes until the task is finished. Otherwise, the scheduling algorithm is considered to be dynamic if the priorities of tasks might change from time to time, although dynamic scheduling is more effective than static scheduling in utilizing the available computational resources. Fixed priority assignment is applied more by industry systems, owing to its efficient implementation, simplicity, and intuitive meaning. For practical purposes, we will focus on the study of static scheduling with fixed priority assignment.

8.3   Schedulability test in the cloud datacenter

The schedulability test predicts temporal behavior of a given task set, and decides whether the deadline constraints will be met at runtime, that is, the given task set can be scheduled. Two main types are

- Sufficient test: All task sets that pass the test can meet their deadlines. However, some task sets that do not pass the test can still be scheduled by the processing resource.

- Exact test: A task set can be scheduled if and only if it passes the test.

In this chapter, we investigate current schedulability tests in terms of design principle, time complexity, and applicable scenario. System designers, who face a tradeoff between test accuracy and overhead, could make a reasonable decision based on the available computational power.

8.3.1   Pseudo-polynomial complexity

An exact schedulability test yields to a sufficient and necessary condition, but it requires high computational complexity [Joseph and Pandya, 1986], even in the simple case where task relative deadlines are equal to periods. Lehoczky [Lehoczky et al., 1989] studied an exact feasibility test with pseudo-polynomial complexity for that RM priority assignment. Based on linear programming, Park [Park et al., 1995] achieved the exact utilization bound without knowledge of exact task computation time. Subsequently, Audsley [Audsley et al., 1993] considered a DM priority assignment and improved Lehoczky's exact feasibility test by searching for worst-case response time in an iterative manner. Lehoczky [Lehoczky, 1990] then proposed a more general feasibility test for arbitrary deadlines. Later, methods for speeding up the analysis of task sets were proposed [Manabe and Aoyagi, 1995, Sjödin and Hansson, 1998, Abdelzaher and Lu, 2001, Bini and Buttazzo, 2002, Chen et al., 2003, Bini and Buttazzo, 2004], but the complexity of the approach always remains pseudo-polynomial in the worst case. Here we present two seminal pseudo-polynomial complexity tests.

Breakdown utilization.   Breakdown utilization was first proposed by Lehoczky [Lehoczky et al., 1989], describing an exact characterization of an RM scheduling algorithm. For a random task set, the computation time scales to the point at which a deadline is first missed. The corresponding set utilization is the breakdown utilization U_n^*. This bound is an exact bound, which provides both sufficient and necessary conditions for a schedulability test. If the utilization of the task set is higher than this bound, no solution exists for scheduling all the tasks on one processor. Otherwise, the task set can be scheduled without missing any deadline. The result seems exciting, but this breakdown utilization changes according to tasks with different periods and computation times. In other words, task set size n is not enough to make a decision, and precise details such as computation time C_i, period T_i for every task should be known in advance.

$$U_n^* = \frac{\sum_{i=1}^n C_i/T_i}{\min_{t \in S_n} \sum_{j=1}^n C_j \lceil t/T_j \rceil / t}$$
$$S_n = kT_j \qquad j = 1, \cdots, n; k = 1, \cdots, \lfloor T_n/T_j \rfloor \tag{8.2}$$

In order to characterize the average behavior, Lehoczky studied the asymptotic behavior of the breakdown utilization when periods and computation times are generated randomly. In particular, U_n^* converges to a constant as the task set size increases, depending only on periods, no longer on computation times. Given task periods generated uniformly in the interval [1,B], breakdown utilization U_n^* converges to.

$$
U_n^* = \begin{cases} 1 & B = 1 \\ \frac{\ln B}{B-1} & 1 < B < 2 \\ \frac{\ln B}{\frac{B}{\lfloor B \rfloor} + \sum_{i=2}^{\lfloor B \rfloor - 1} 1/i} & B \geq 2 \end{cases} \tag{8.3}
$$

and the rate of convergence is $O(\sqrt{n})$.

In addition, the function of U_n^* with respect to B first decreases and then increases as B grows from one to infinity, bottoming at $B = 2$, which is in agreement with Liu's result. For uniformly distributed tasks, 0.88 is a reasonable approximation for the breakdown utilization bound, which is much larger than Liu's sufficient bound of 0.69.

Response time analysis. Breakdown utilization has a strict restriction that the deadline of a task must equal the period. For tasks with deadlines no more than periods, DM is the optimal priority assignment [Leung and Whitehead, 1982]. Audsley proposed a method to estimate the actual worst response time for each task, so the schedulability test turns out to be a trivial comparison of each task's response time and its deadline.

Response time is the period between task submission and execution completion. The worst response time R_i for a task i equals the sum of its computation time C_i and the worst interference I_i. Interference is defined as the preemption time of higher priority tasks $(j < i)$, and is given by the sum of $\left\lceil \frac{R_i}{T_j} \right\rceil C_j$.

$$
R_i = C_i + \sum_{\forall j < i} \left\lceil \frac{R_i}{T_j} \right\rceil C_j \tag{8.4}
$$

R_i can be calculated by asymptotic iteration.

$$
R_i^{n+1} = C_i + \sum_{\forall j < i} \left\lceil \frac{R_i^n}{T_j} \right\rceil C_j \tag{8.5}
$$

R_i^n is the nth iteration. The iteration begins at $R_i^0 = 0$, and ends at $R_i^{n+1} = R_i^n$. If R_i^n reaches D_i before termination of convergence, iteration also halts, that is to say, the task set is not schedulable. This analysis intends to predict the worst interference that a task can suffer from higher priority tasks. Since the prediction formulation does not refer to any priority assignment strategy, it is effective for both RM and DM approaches.

8.3.2   Polynomial complexity

Response time analysis (RTA) is a popular method for schedulability analysis of real-time tasks. Many efforts in the simplification of RTA have been made by reducing the number of iterations [Sjödin and Hansson, 1998, Bril et al., 2003, Lu et al.,

2006]. Although some of them can shorten the runtime with a saving of 26–33% calculation [Lu et al., 2006], all currently known algorithms still take a runtime pseudo-polynomial in the representation of the task system. Besides that, approximation is then applied to further reduce the time complexity of an exact schedulability test.

Fisher [Fisher and Baruah, 2006] derived a fully polynomial time approximation scheme of the RTA. This scheme accepts two inputs, the specifications of a task system and a constant $\epsilon \in [0, 1]$, to examine feasibility tests. If the test returns feasible, the task set is guaranteed to be scheduled on the processor for which it has been specified. If the test returns unfeasible, the task set is guaranteed to be unscheduled on a slower processor, the computing capacity of which is in $(1 - \epsilon)$ proportion to the specified processor.

The number of iterations of interference calculation is limited to a constant k, where $k = \lceil 1/\epsilon \rceil - 1$. So the approximated value of I_i is

$$\tilde{I}_i = \begin{cases} \left\lceil \frac{t}{T_i} \right\rceil C_i & t \le (k-1)T_i \\ C_i + \frac{t}{T_i}C_i & t > (k-1)T_i \end{cases} \tag{8.6}$$

Therefore, the worst response time $\tilde{R}_i$ is calculated in $O(n^2 k)$ time complexity.

$$\tilde{R}_i = C_i + \sum_{\forall j < i} \tilde{I}_i \tag{8.7}$$

In addition, Bini [Bini and Baruah, 2007] derived an upper bound on the response times in polynomial time. The worst response time R_i is bounded by R_i^{ub} as

$$R_i \le \frac{C_i + \sum_{j=1}^{i-1} C_j(1 - U_j)}{1 - \sum_{j=1}^{i-1} U_j} = R_i^{ub} \tag{8.8}$$

The time complexity of computing the response time upper bound R_i^{ub} is $O(i)$, and the complexity of computing the bound for all the tasks is $O(n^2)$.

More polynomial complexity tests can apply the utilization bounds presented in the previous chapter. For example, Han [Han, 1998] suggested modifying the task set with smaller, but harmonic, periods using an algorithm with $O(n^2 \log n)$ complexity. Chen [Chen et al., 2003] investigated an algorithm with $O(n^3)$ complexity that obtains an exact bound under the condition where periods and computation times are integers. Lauzac limited period relations, and improved schedulability within a $O(n \log n)$ time complexity.

Generally speaking, all polynomial complexity tests are only sufficient, not necessary. The time complexity for exact tests is always NP-hard for non-trivial computational models [Sha et al., 2004]. Less complexity is always achieved at the cost of less accuracy.

8.3.3 Constant complexity

The constant complexity tests apply the simplest bound, such as the classic bound [Liu and Layland, 1973] or the hyperbolic bound [Bini et al., 2003]. Both of these tests are in $O(1)$ time complexity, so they are easily implemented and fast enough for on-line schedulability tests. As long as the utilization of a given task set is under this bound, all tasks can be scheduled for sure. One shortcoming is that the two bounds are only suitable for RM approach. In order to determine a concise schedulable condition like Liu's result, Peng [Peng and Shin, 1993] proposed a concept of system hazard to check whether assigned tasks miss their deadlines, and computed the lowest upper bound of DM algorithm. The calculation of DM bound can be finished in $O(1)$ time complexity.

Recently, another schedulability test with $O(1)$ constant complexity has been developed by Masrur [Masrur et al., 2010, Masrur and Chakraborty, 2011]. This test calculates an upper bound of the worst response time considering all accepted tasks, and is different from all mentioned tests based on system utilization. If this upper bound does not exceed the respective deadlines, all tasks can be scheduled under DM. However, the comparison with other bound-based tests remains unfinished by the authors.

8.4   Utilization bounds for schedulability testing

Utilization bound $\widehat{U}$ provides a simple and practical way to test the schedulability of a real-time task set. If the system utilization of a given task set $\sum u_i$ is lower than the bound $\widehat{U}$, the task set can be scheduled by a processing resource. Although the bound is only sufficient, not necessary, it is widely used in industry, because it is easily implemented and fast enough for on-line tests. The simplest bound is decided by the number of tasks in a task set. To raise the system utilization bound, strict constrains are relaxed by subsequent researchers. The more information on the task set included, the better the utilization bound obtained. In this section, we revisit the development of the utilization bound.

8.4.1   Classical bound

In 1973, Liu [Liu and Layland, 1973] proposed a Rate Monotonic (RM) scheduling algorithm for preemptive periodic tasks on a uniprocessor in a hard real-time system, which played seminal roles in the development of real-time scheduling research. The RM algorithm assigns priorities to tasks inversely proportional to their periods. Liu proved an RM algorithm is the optimal fixed priority assignment, and derived the lowest upper bound from the worst case of system utilization by arbitrary task set, that is

$$\widehat{U} = n(2^{1/n} - 1) \qquad (8.9)$$

This bound decreases monotonically from 0.83 to 0.69 when n approaches infinity. As long as the utilization of a given task set is beneath this bound, schedulability is guaranteed. However, this bound is only sufficient, not necessary. Many task sets with utilization higher than this bound can still be scheduled. This phenomenon implies that the processing resource is underutilized. The desire to improve the system utilization leads to research on a more precise bound.

8.4.2 Closer periods

Burchard [Burchard et al., 1994] found an increasing utilization if all periods in a task set have values that are close to each other. For a set of n tasks, Burchard introduced two parameters $S_i = \log_2 T_i - \lfloor \log_2 T_i \rfloor$ and $\beta = \max S_i - \min S_i$. The least upper bound of processor utilization is

$$\widehat{U} = \begin{cases} (n-1)(2^{\beta/n-1} - 1) + 2^{1-\beta} - 1 & \beta < 1 - 1/n \\ n(2^{1/n} - 1) & \beta \geq 1 - 1/n \end{cases} \tag{8.10}$$

Higher utilization can be obtained if task periods satisfy a certain constraint $\beta < 1 - 1/n$. The disadvantage is that more calculation is needed, such as searching for the explicit task periods.

8.4.3 Harmonic chains

Appropriate choice of task periods guarantees high utilization, especially when task periods are harmonic. Sha proved that schedulability is guaranteed up to 100% utilization with harnomic periods [Sha and Goodenough, 1990]. The limitation of periods hedges the practice in the application domain. Kuo [Kuo and Mok, 1991] generalized this result by grouping tasks in serveral harmonic chains. Every harmonic chain is a list of numbers in which every number divides every number after it. If there are k harmonic chains, clearly $k \leq n$, the least upper bound to processor utilization is

$$\widehat{U} = k(2^{1/k} - 1) \tag{8.11}$$

A better bound is obtained by applying period parameters. However, determining the number of harmonic chains for a given task set also increases the time complexity.

Chen [Chen et al., 2003] investigated an exact bound that can be derived exhaustively under the condition where periods and computation times are integers. An algorithm with $O(n^3)$ complexity is presented and performs better than a harmonic bound. He also proposed another algorithm, which yields an exact bound with exponential complexity.

8.4.4 Hyperbolic bound

Bini [Bini et al., 2003] proposed a schedulability condition similar to a utilization bound. This condition does not depend on the number of tasks. The schedulability test using Bini's result has the same complexity as using Liu's bound, but is less pessimistic. For a set of n tasks with fixed priority order, where each task is characterized by a single utilization u_i, the task set is schedulable if

$$\prod (u_i + 1) \leq 2 \tag{8.12}$$

This result can also be integrated into the method of harmonic chains.

8.5   Real-time task scheduling with MapReduce

We concentrate on looking for a utilization bound on a MapReduce cluster for on-line schedulability analysis, because exact tests are nearly intractable in real-time systems. Their time complexity is NP-hard for these non-trivial computational models [Sha et al., 2004], which is not acceptable for an on-line test.

The improvement of bound is achieved by introducing practical requirements of applications. When periodic tasks are executed on a MapReduce cluster, the combination of sequential computing and parallel computing impacts real-time scheduling. In the next section, we analyze how the segmentation between Map and Reduce influences cluster utilization.

8.5.1   System model

Assume a task set $\Gamma = (\tau_1, \tau_2, \cdots, \tau_n)$ including n periodic tasks on a MapReduce cluster. Without losing generality, we let $T_1 < T_2 < \cdots < T_n$. In RM scheduling, task with higher request rate has higher priority, so task τ_1 with shortest period has highest priority, while the last τ_n has the lowest. All tasks are independent, that is, have no precedence relationship. Besides that, all tasks are fully preemptive, and the overhead of preemptive is negligible.

MapReduce solves distributable problems using a large number of computers, collectively referred to as a cluster, with certain computing capability. One task is partitioned into n_m Maps and n_r Reduces. The numbers of n_m and n_r are not fixed, varying from one task to another. Maps performed in parallel finish in a certain time M_i, which means total time required to complete n_m Map operations. Total time spent on n_r Reduces is execution time R_i. For simplification, we assume R_i is in proportion to M_i, and $\alpha = R_i/M_i$ is introduced to express the ratio between the two operations. Here we simply let all tasks use the same α. The whole computation time for task τ_i is

$$C_i = M_i + R_i = M_i + \alpha M_i = \frac{1}{\alpha} R_i + R_i \qquad (8.13)$$

One remarkable character of MapReduce is that no Reduce operation can be submitted until all Map operations finish, so M_i and R_i have innate temporal sequence and share no overlap. In the following context, we use Map/Reduce to signify the whole executing process of Map/Reduce operations.

As in the former assumption, request of each instance occurs when a new period begins, so the Map request is consistent with the request of the whole task. The moment when a Reduce request is submitted makes a huge impact on cluster utilization. If Reduce always executes as soon as Map finishes, the two stages of Map and

Reduce are continuous. Hence the task can be considered as a general case without segmentation, the bound of which is the famous Liu's bound. If Reduce does not make its request in a hurry, this tradeoff can be beneficial to cluster utilization by making better use of spare time. We introduce parameter $\beta = T_{R_i}/T_{M_i}$ to reveal the segmentation ratio. The same β is applied for all tasks in task set Γ. Clearly,

$$T_i = T_{M_i} + T_{R_i} = T_{M_i} + \beta T_{M_i} = \frac{1}{\beta}T_{R_i} + T_{R_i} \qquad (8.14)$$

Utilization u_i is the ratio of computation time to its period $u_i = C_i/T_i$. System utilization U is the sum of utilization for all the tasks in the task set.

$$U = \sum_{i=1}^{n} u_i = \frac{M_1 + R_1}{T_1} + \frac{M_2 + R_2}{T_2} + \cdots + \frac{M_n + R_n}{T_n} \qquad (8.15)$$

8.5.2 MapReduce segmentation

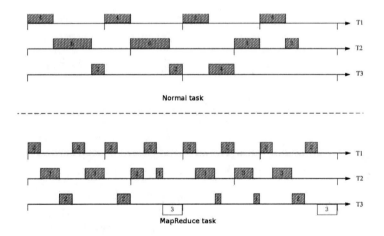

FIGURE 8.2: Comparison of a normal task and a MapReduce task.

Seen from the above system model, there is a natural segmentation between Map and Reduce. For a MapReduce task, a delay might exist during the whole execution time, in contrast with a normal task executed in one go from beginning to end. How does this characteristic impact on the schedulability performance? We take Figure 8.2 for example, to give an intuitive idea.

There is a task set $\Gamma = (\tau_1, \tau_2, \tau_3)$, in which $C_1 = 4, C_2 = 6$. Because $T_1(12) < T_2(16) < T_3(24)$, task τ_1 has the highest execution priority, while task τ_3 has the lowest.

First, we analyze the case of normal tasks. In order to fully utilize a cluster, the computation time C_3 of task τ_3 is no more than 4. Otherwise, the cluster fails in scheduling these three tasks simultaneously. In this case, the system utilization is $U_{Normal} = \frac{C_1}{T_1} + \frac{C_2}{T_2} + \frac{C_3}{T_3} = 0.875$

Next, we consider the case of MapReduce tasks with $\alpha = 1$. Computation time C_3 can be increased to 7 from 4, without changing C_1 and C_2. The system utilization is then $U_{MapReduce} = \frac{M_1+R_1}{T_1} + \frac{M_2+R_2}{T_2} + \frac{M_1+R_3}{T_3} = 1$.

Therefore, the system utilization augments owing to the segmentation between Map and Reduce. Quantitative analysis of exact augmentation is presented in the next section.

8.5.3   Worst pattern for a schedulable task set

To begin with, let us review the concept of critical instant theorem proposed by Liu [Liu and Layland, 1973].

THEOREM 8.1

A critical instant of a task is the moment at which the task makes a request which has the largest response time. It happens whenever the task is requested simultaneously with all higher priority tasks.

The concept implies the worst case occurs when all the tasks start to make requests at the same time. Therefore, the offsets of all tasks are set to zero, that is, $O_i = 0$. In order to decide whether a task set is schedulable, we check if and only if the first request of each task is met in their first period when all tasks begin simultaneously.

In this section, T_{M_i} and T_{R_i} are treated as relative deadlines for Map and Reduce, respectively. Map M_i instantiates at the beginning of a new period, and must be finished before T_{M_i}. At the moment T_{M_i}, Reduce R_i makes the request, and its execution lasts for T_{R_i} at most. In this assumption, how does the cluster bound change according to the given latency T_{M_i}? In order to get the lowest utilization, we find out the worst pattern for a schedulable task set on the MapReduce cluster. Lemma 8.1 depicts the worst pattern for a schedulable task set that fully utilizes a MapReduce cluster.

LEMMA 8.1

For a task set $\Gamma = (\tau_1, \tau_2, \cdots, \tau_n)$ with fixed priority assignment where $T_n > T_{n-1} > \cdots > T_2 > T_1$, if the relative deadline of Reduce is not longer than Map ($\beta \leq 1$), the worst pattern ensuring all tasks to be scheduled is

$$M_1 = T_2 - T_1$$
$$M_2 = T_3 - T_2$$
$$\vdots$$
$$M_{n-1} = \frac{1}{1+\beta}T_n - T_{n-1}$$
$$M_n = (2+\alpha)T_1 - \frac{1+\alpha}{1+\beta}T_n$$

PROOF   Suppose a task set is fully utilizing a MapReduce cluster. Fully utilizing has two meanings. The first implies that a task set can be scheduled on a cluster, and the second shows that no improvement can be made in terms of cluster utilization. Each task τ_i in task set Γ is defined by a triple $< M_i, R_i, T_i >$, or equally $< M_i, \alpha, T_i >$ considering $\alpha = R_i/M_i$.

In order to analyze the period relationship between two neighboring tasks with the most adjacent priorities, we assume that

$$M_1 = T_2 - \left\lfloor \frac{T_2}{T_1} \right\rfloor \cdot T_1 + \epsilon \tag{8.16}$$

Where ϵ is a real number. We reduce Map runtime M_1 with ϵ when $\epsilon > 0$. In order to maintain the full processor utilization, M_2^a is improved with the amount of ϵ.

$$
\begin{array}{ll}
M_1^a = T_2 - \left\lfloor \frac{T_2}{T_1} \right\rfloor \cdot T_1 & T_1^a = T_1 \\
M_2^a = M_2 + \epsilon & T_2^a = T_2 \\
\vdots & \vdots \\
M_n^a = M_n & T_n^a = T_n
\end{array} \tag{8.17}
$$

Through this adjustment cluster utilization U^a is consequently smaller than original utilization U, because

$$U - U^a = \epsilon(1+\alpha)(\frac{1}{T_1} - \frac{1}{T_2}) > 0 \tag{8.18}$$

Although the two task sets fully utilize the cluster, the latter has a low cluster utilization. As a result, the new pattern is worse than the former one.

On the contrary, when $\epsilon < 0$, M_2 gets longer to fully use the cluster as

$$
\begin{array}{ll}
M_1^b = T_2 - \left\lfloor \frac{T_2}{T_1} \right\rfloor \cdot T_1 & T_1^b = T_1 \\
M_2^b = M_2 + \left\lceil \frac{T_2}{T_1} \right\rceil \cdot \epsilon & T_2^b = T_2 \\
\vdots & \vdots \\
M_n^b = M_n & T_n^b = T_n
\end{array} \tag{8.19}
$$

The corresponding utilization U^b decreases again, owing to

$$U - U^b = \epsilon(1+\alpha)(\frac{1}{T_1} - \left\lceil \frac{T_2}{T_1} \right\rceil \frac{1}{T_2}) > 0 \tag{8.20}$$

The worst pattern of a task set makes cluster utilization reach minimum, as long as ϵ approaches zero. The following analysis is based on the condition $\epsilon = 0$.

Next, the period T_1 enlarges $\left\lfloor \frac{T_2}{T_1} \right\rfloor$ times as

$$
\begin{aligned}
M_1^c &= T_2 - \left\lfloor \tfrac{T_2}{T_1} \right\rfloor \cdot T_1 & T_1^c &= \left\lfloor \tfrac{T_2}{T_1} \right\rfloor \cdot T_1 \\
M_2^c &= M_2 + (\left\lfloor \tfrac{T_2}{T_1} \right\rfloor - 1)(T_2 - \left\lfloor \tfrac{T_2}{T_1} \right\rfloor T_1) & T_2^c &= T_2 \\
&\ \vdots & &\ \vdots \\
M_n^c &= M_n & T_n^c &= T_n
\end{aligned}
\tag{8.21}
$$

Compare new utilization U^c with U

$$
U - U^c = (1 + \alpha)(1 - 1/\left\lfloor \frac{T_2}{T_1} \right\rfloor)(\frac{T_2}{T_1}) \geq 0 \tag{8.22}
$$

This revision further pulls down the cluster utilization, which leads us to conclude that closer periods degrade the system utilization. If we try to search for the worst pattern, the smallest value of $\left\lfloor \frac{T_2^c}{T_1^c} \right\rfloor$ should be taken. Hence the worst case happens when $\left\lfloor \frac{T_2^c}{T_1^c} \right\rfloor = 1$, in other words, $T_2 < 2T_1$.

To sum up, we have

$$
M_1 = T_2 - T_1 \tag{8.23}
$$

Using similar methods, we obtain more results about the period relationship between two adjacent tasks.

$$
M_i = T_{i+1} - T_i, \qquad i = 2, 3, \cdots, n - 2 \tag{8.24}
$$

For the purpose of analyzing the relationship between T_{n-1} and T_n, we construct a new task set by halving the period T_{n-1}. The periods of other tasks keep the same $T_1, T_2, \cdots T_{n-2}, T_n$. To avoid any waste, Map execution time M_{n-1} is transferred from τ_{n-1} to τ_n.

$$
\begin{aligned}
M_1^d &= M_1 & T_1^d &= T_1 \\
M_2^d &= M_2 & T_2^d &= T_2 \\
&\ \vdots & &\ \vdots \\
M_{n-1}^d &= 0 & T_{n-1}^d &= T_{n-1}/2 \\
M_n^d &= M_n + M_{n-1} & T_n^d &= T_n
\end{aligned}
\tag{8.25}
$$

A lower utilization U^d is achieved, comparing with old U

$$
U - U^d = M_{n-1}(\frac{1}{T_{n-1}} - \frac{1}{T_n}) > 0 \tag{8.26}
$$

The task set is resorted according to the length of period assuring $T_n > T_{n-1} > \cdots > T_2 > T_1$. The new pattern further decreases utilization under the condition that $T_{n-1}^d < \frac{1}{1+\beta}T_n^d$, owing to $T_{n-1}^d \leq \frac{1}{1+\beta} \cdot 2T_{n-1}^d = \frac{1}{1+\beta}T_{n-1} < \frac{1}{1+\beta}T_n = \frac{1}{1+\beta}T_n^d$.

Map M_{n-1} is obtained

$$M_{n-1} = \frac{1}{1+\beta}T_n - T_{n-1} \tag{8.27}$$

Time left for Map execution M_n is

$$M_n = \frac{1}{1+\beta}T_n - \sum_{i=1}^{n-1} C_i - \sum_{i=1}^{n-1} M_i = (2+\alpha)T_1 - \frac{1+\alpha}{1+\beta}T_n \tag{8.28}$$

$\square$

The above worst pattern stands for the most pessimistic situation where the least utilization can be calculated. Under the condition given by Lemma 2, schedulable upper bound on MapReduce cluster is derived.

THEOREM 8.2
   For a task set $\Gamma = (\tau_1, \tau_2, \cdots, \tau_n)$ with a fixed priority assignment where $T_n > T_{n-1} > \cdots > T_2 > T_1$, if the length of reduce is not longer than map ($\beta \leq 1$), the schedulable upper bound of cluster utilization is $U = (1 + \alpha)\left\{n[(\frac{2+\alpha}{1+\beta})^{1/n} - 1] + \frac{\beta-\alpha}{1+\beta}\right\}$.

PROOF   To simplify the notation, parameters $\gamma_1, \gamma_2, \cdots, \gamma_n$ are introduced

$$T_i = \gamma_i T_n \qquad i = 1, 2, \cdots, n-1 \tag{8.29}$$

Computation time of n tasks is expressed as

$$\begin{aligned} C_i &= (1+\alpha)(\gamma_{i+1}T_n - \gamma_i T_n) \qquad i = 1, 2, \cdots, n-2 \\ C_{n-1} &= (1+\alpha)(\frac{1}{1+\beta}T_n - \gamma_{n-1}T_n) \\ C_n &= (1+\alpha)[(2+\alpha)\gamma_1 T_n - \frac{1+\alpha}{1+\beta}T_n] \end{aligned} \tag{8.30}$$

Which gives the cluster utilization U

$$U = (1+\alpha)[\sum_{i=1}^{n-2} \frac{\gamma_{i+1} - \gamma_i}{\gamma_i} + \frac{\frac{1}{1+\beta} - \gamma_{n-1}}{\gamma_{n-1}} + (2+\alpha)\gamma_1 - \frac{1+\alpha}{1+\beta}] \tag{8.31}$$

In order to compute the minimum value of U, we set the first order partial derivative of function U with respect to variable γ_i to zero

$$\frac{\partial U}{\partial \gamma_i} = 0 \qquad i = 1, 2, \cdots, n-1 \tag{8.32}$$

For variable γ_i, we get the equation

$$\begin{aligned}
\gamma_1^2 &= \frac{1}{2+\alpha}\gamma_2 \\
\gamma_i^2 &= \gamma_{i-1}\gamma_{i+1} \qquad i = 2, \cdots, n-1
\end{aligned} \tag{8.33}$$

The general expression of γ_i is

$$\gamma_i = \frac{1}{2+\alpha}\left(\frac{2+\alpha}{1+\beta}\right)^{i/n} \qquad i = 1, 2, \cdots, n-1 \tag{8.34}$$

By substituting general value of γ_i into U, the least cluster utilization is achieved as

$$U = (1+\alpha)\left\{n\left[\left(\frac{2+\alpha}{1+\beta}\right)^{1/n} - 1\right] + \frac{\beta-\alpha}{1+\beta}\right\} \tag{8.35}$$

$$\square$$

Moreover, a symmetric utilization bound is easily deduced using a similar method as Theorem 2. If the length of reduce is longer than map ($\beta > 1$), the schedulable upper bound of cluster utilization is

$$U = \left(1+\frac{1}{\alpha}\right)\left\{n\left[\left(\frac{\beta+2\alpha\beta}{\alpha+\alpha\beta}\right)^{1/n} - 1\right] + \frac{\alpha-\beta}{\alpha+\alpha\beta}\right\} \tag{8.36}$$

On a real MapReduce cluster, numerous tasks are executed concurrently, so the number n is typically very large. Therefore, for all practical purposes, we are most interested in the cluster utilization as $n \to \infty$. When n is infinite, the limit of U is

$$U_\infty = \lim_{n\to\infty} U = \begin{cases} (1+\alpha)[\ln(\frac{2+\alpha}{1+\beta}) + \frac{\beta-\alpha}{1+\beta}] & \beta < 1 \\ (1+\frac{1}{\alpha})[\ln(\frac{\beta+2\alpha\beta}{\alpha+\alpha\beta}) + \frac{\alpha-\beta}{\alpha+\alpha\beta}] & \beta \geq 1 \end{cases} \tag{8.37}$$

Figure 8.3 outlines the fluctuation of the utilization bound with respect to α and β, where α shows the proportion of execution time between Map and Reduce and β illustrates the ratio between two relative deadlines. Seen from Figure 8.3, the bound is a symmetrical plane on the axis $\alpha = \beta$. It implies that the value of α and β should be harmonious, that is, difference between α and β can not be too dramatic. Easily understood, if a long Map ($\alpha < 1$) is given a short relative deadline ($\beta > 1$), it is impossible to schedule all the tasks before periods expire. That is why the cluster utilization dips to zero when assignment of the two variables goes in opposite directions. If α and β are given reasonable values in advance and task sets can be scheduled on the MapReduce cluster, utilization bound is a concave function with respect to α and β. Figure 8.4 is drawn when $\alpha = \beta$. The bound rises steadily due to segmentation of Map and Reduce. When β approaches zero, the least cluster utilization is near 0.7. The amount of utilization rises as β goes up until $\beta = 1$, peaking at 0.81, which is also a global maximum point. After that, the cluster bound declines slowly to 0.7 again, when β increases to infinity. Notice that Liu's utilization bound $U = n(2^{\frac{1}{n}} - 1)$ can be represented in a task set with $\beta = 0$ or $\beta = \infty$. $\beta = 0$

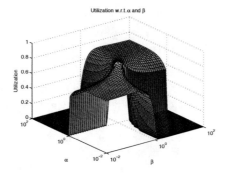

FIGURE 8.3: Utilization bound.

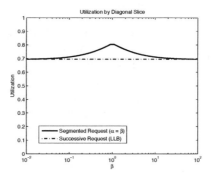

FIGURE 8.4: Optimal utilization.

is an extreme case where the time spent on Reduce is negligible, so Map execution time stands for the whole computation time. The case of $\beta = \infty$ implies that Reduce execution occupies the whole computation time. Therefore, our new bound is a general expression of Liu's bound, only if a special value of β is assigned in these functions.

Our result improves on Liu's work. This augmentation comes from the flexibility of MapReduce. The execution of Map operations should be first promised, because Reduces need to collect all the output of Maps. However, the moment when Reduce makes a request changes the final cluster utilization. If Reduce waits in a reasonable period and hands over the cluster to a more pressing task with lower priority, it is possible to achieve more dynamic allocation and a higher utilization bound than in the case in which no segmentation exists between Map and Reduce.

8.6 Reliability indication methods

Since the real-time requirement is a significant QoS criterion of cloud service provision, an on-line schedulability test is necessary. This test can determine whether an arriving application is accepted or not, so it can guarantee the system stability.

Several popular schedulability tests in real-time computing are presented, among which a comprehensive analysis is built, in terms of balance between time complexity and acceptance ratio. Some schedulability tests yield to exact conditions to achieve the maximum system utilization, but the time complexity is not acceptable for an on-line test. Some of them applying sufficient conditions might somewhat underutilize the cluster, but can be finished quickly, in predictable running time. Herein, we focus on the tests with constant-time complexity, which is more suitable for on-line guarantee in cloud context.

Although a number of schedulability tests have been studied, they are incomparable if the determination conditions are different. In order to maintain high system utilization, the problem of choosing a reliable test attracts our attention. Typically, simulation can give an intuitive answer, but the result always depends on the way random parameters are generated and the number of experiments. Therefore, we introduce a concept of test reliability to evaluate the probability that a random task set can pass a given schedulability test, and we define an indicator to express the test reliability. The larger the probability, the more reliable the test is. From the point of view of a system, a test with high reliability can guarantee high system utilization.

8.6.1 Reliability indicator

The effectiveness of a sufficient schedulability test can be measured by the accepted ratio of task sets. The larger the ratio is, the more reliable the test is. One typical method to calculate acceptable ratio is Monte Carlo simulation, in which a large number of synthetic task sets need to be generated with random parameters. However, almost all measurements are made with some intrinsic errors. If the method of generating parameters is biased, unreasonable conclusions might be deduced due to the different hypotheses between simulations and actual working conditions. For these reasons, a probability method is used to indicate the likelihood of an acceptable ratio.

Note that this accepted ratio is different from the similar concept in previous researches [Bini and Buttazzo, 2004]. The denominator of this ratio is the total number of participated tests, rather than the number of feasible ones. Such an adjustment makes our analysis much easier, because finding out all feasibly schedulable task sets in an exact test is extremely time consuming. Another advantage is that a simple UUniform algorithm turns out to be practical in our simulation, which does not work for original test of accepted ratio, owing to a huge number of iterations [Bini et al., 2003].

Without loss of generality, we suppose that task utilization u_i is uniformly distributed with mean value $1/2$ and variance $1/12$. Two probability distributions will be calculated in the following context.

(1) $X = \sum_{i=1}^{n} u_i$

X is the sum of n independent u_i, and the Probability Density Function (PDF) of X is

$$\mathcal{F}_{PDF}(X) = \frac{1}{(n-1)!} \sum_{k=0}^{\lfloor U \rfloor} (-1)^k \binom{n}{k} (U-k)^{n-1} \quad U \in [0, n] \tag{8.38}$$

Therefore, U has mean value $n/2$ and variance $n/12$. Its Cumulative Distribution Function (CDF) is

$$\mathcal{F}_{CDF}(X) = \frac{1}{n!} \sum_{k=0}^{\lfloor U \rfloor} (-1)^k \binom{n}{k} (U-k)^n \quad U \in [0, n] \tag{8.39}$$

More generally, for a sequence of independent and identically distributed random variables u_i with expected values μ and variances σ^2, the central limit theorem asserts that for large n, the distribution of the sum X is approximately normal with mean $n\mu$ and variance $n\sigma^2$.

$$X \to \mathcal{N}(\frac{n}{2}, \frac{n}{12}) \tag{8.40}$$

(2) $Y = \sum_{i=1}^{n} 2u_i/(1 + u_i)$

An intermediate variable $y_i = 1/(1 + u_i)$ is introduced, and its PDF is expressed as

$$\mathcal{G}_{PDF}(y_i) = \frac{1}{y_i^2} \quad y_i \in [\frac{1}{2}, 1] \tag{8.41}$$

Mean and variance of y_i are

$$E(y_i) = \int_{\frac{1}{2}}^{1} y_i g(y_i) \mathrm{d}y_i = \ln 2 \tag{8.42}$$

$$D(y_i) = E(y_i^2) - [E(y_i)]^2 = \frac{1}{2} - (\ln 2)^2 \tag{8.43}$$

With y_i, we obtain

$$Y = \sum_{i=1}^{n} \frac{2u_i}{1 + u_i} = \sum_{i=1}^{n} 2(1 - y_i) \tag{8.44}$$

Y is approximated by a normal distribution as

$$Y \to \mathcal{N}[2n(1 - \ln 2), 4n(\frac{1}{2} - (\ln 2)^2)] \tag{8.45}$$

We define reliability indicator w as

$$w = \frac{x - \mu}{\sigma} \tag{8.46}$$

For a generic normal random variable with mean μ and variance σ^2, the CDF is $F(x) = \Phi(\frac{x-\mu}{\sigma})$, in which $\Phi(x)$ is the standard normal distribution. Since the CDF of $\Phi(w)$ is a monotone increasing function with respect to w, w can indicate the probability that a random task set passes a given examination. The higher the probability obtained, the better the examination is. Therefore, different schedulability tests can be compared by a reliability indicator. The test with a large value of w is more reliable than that with a small value.

8.6.2   Schedulability test conditions

Liu's RM condition.   RM scheduling is an optimum static algorithm [Liu and Layland, 1973]. If RM can not make a task set schedulable on a cluster, no other rules can succeed in scheduling. RM algorithm is only suitable for the cases in which a task period exactly equals its deadline. Liu proposed a concept of system utilization U as a sufficient condition for a schedulability test. The subscripts l, p and m represent the work of Liu, Peng and Masrur detailed in the following content, respectively.

THEOREM 8.3
For a set of n tasks with fixed utilization $u_1, u_2, \cdots, u_n$, there exists a feasible algorithm ensuring all tasks can be scheduled on a cluster if

$$U_l = \sum_{i=1}^{n} u_i \leq n(2^{1/n} - 1) \tag{8.47}$$

Peng's DM condition.   Deadline replaces period as the new determinant when deadline does not equal period. Peng [Peng and Shin, 1993] modified the system utilization U_p for DM algorithm by introducing system hazard $\theta = D_i/T_i, 1 \leq i \leq n$.

THEOREM 8.4
For a set of n tasks with fixed utilization $u_1, u_2, \cdots, u_n$, there exists a feasible algorithm ensuring all tasks can be scheduled on a cluster if

$$U_p = \sum_{i=1}^{n} u_i \leq \begin{cases} \theta & \theta \in [0, 0.5) \\ n[(2\theta)^{1/n} - 1] + 1 - \theta & \theta \in [0.5, 1] \end{cases} \tag{8.48}$$

Masrur's DM condition.   Masrur [Masrur et al., 2010] also studied a set of tasks with deadline no longer than period, and proposed a load condition to test whether a task set is schedulable on a cluster.

THEOREM 8.5
For a set of n tasks with fixed utilization $u_1, u_2, \cdots, u_n$, there exists a feasible

algorithm ensuring all tasks can be scheduled on a cluster if

$$\sum_{i=1}^{n} \max\left(\frac{u_i}{\theta}, \frac{2u_i}{1+u_i}\right) \leq 1 \tag{8.49}$$

Masrur's condition contains a maximum operator. For the sake of simplicity, we replace the max by introducing two parameters $u_l = (1 + \min u_i)/2$ and $u_h = (1 + \max u_i)/2$. There are m tasks ($m \leq n$) satisfy that u_i/θ is larger than $2u_i/(1+u_i)$. Then Masrur's condition is decomposed to

$$U_m = \begin{cases} \frac{1}{\theta} \sum_{i=1}^{n} u_i \leq 1 & \theta \in [0, u_l) \\ \frac{1}{\theta} \sum_{i=i}^{m} u_i + \sum_{j=1}^{n-m} \frac{2u_j}{1+u_j} \leq 1 & \theta \in [u_l, u_h) \\ \sum_{i=1}^{n} \frac{2u_i}{1+u_i} \leq 1 & \theta \in [u_h, 1] \end{cases} \tag{8.50}$$

8.6.3 Comparison of rate monotonic conditions

When deadline equals period, we have two RM sufficient conditions for schedulability test.

Based on Liu's condition (8.47) and (8.40), we get $\mu = n/2, \sigma = \sqrt{n/12}$ and $x = n(2^{1/n} - 1)$, hence the reliability indicator of Liu's condition is

$$w_l = \frac{x - \mu}{\sigma} = \frac{n(2^{1/n} - 1) - \frac{n}{2}}{\sqrt{\frac{n}{12}}} \tag{8.51}$$

According to Masrur's load test, we obtain

$$U_m = \sum_{i=1}^{n} \frac{2u_i}{1+u_i} \leq 1 \tag{8.52}$$

From (8.45), $\mu = 2n(1 - \ln 2), \sigma = \sqrt{4n(\frac{1}{2} - (\ln 2)^2)}$ and $x = 1$, so the reliability indicator of Masrur's condition is

$$w_m = \frac{x - \mu}{\sigma} = \frac{1 - 2n(1 - \ln 2)}{\sqrt{4n(\frac{1}{2} - (\ln 2)^2)}} \tag{8.53}$$

The comparison between the two reliability indicators has been plotted in Figure 8.5. Notice that w_l is always larger than w_m, which implies that the schedulability test using Liu's condition is more reliable than that using Masrur's condition. This comparison result can be more intuitive when we focus on the accepted probability of the two tests. Figure 8.6 shows the comparison of accepted probability with different numbers of tasks, ranging from 2 to 20. Masrur's test is more pessimistic, because an arbitrary task set has lower probability of succeeding in Masrur's test than in Liu's test. The difference between two accepted ratios diminishes as the number of tasks augments. When the number reaches a certain value, the reliability of the two tests is

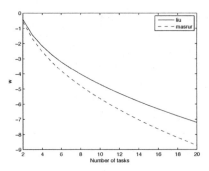

FIGURE 8.5: Comparison of reliability indicators.

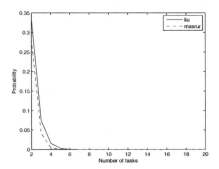

FIGURE 8.6: Comparison of accepted probabilities.

nearly the same. In Figure 8.5 and Figure 8.6, the gap between two reliability indicators increases when the gap between acceptable possibility is reduced. Hence, the reliability indicator can only show relative difference of test reliability, rather than absolute performance.

8.6.4   Comparison of deadline monotonic conditions

Next, the limitation that the deadline exactly equals the period is broken. In DM scheduling, we also analyze two schedulable conditions when the deadline is not longer than period.

According to Peng's condition (8.48) and (8.40), we obtain $\mu = n/2$, $\sigma = \sqrt{n/12}$. The reliability indicator is

$$w_p = \begin{cases} \frac{\theta - \frac{n}{2}}{\sqrt{\frac{n}{12}}} & \theta \in [0, 0.5) \\ \frac{n[(2\theta)^{1/n} - 1] + 1 - \theta - \frac{n}{2}}{\sqrt{\frac{n}{12}}} & \theta \in [0.5, 1] \end{cases} \tag{8.54}$$

w_p is a function of two variables of n and θ, and its gradient vector is

$$\nabla w_p(n, \theta) = \left(\frac{\partial w_p}{\partial n}, \frac{\partial w_p}{\partial \theta} \right) \quad (8.55)$$

The gradient vector implies: (a) $\frac{\partial w_p}{\partial n} < 0$ means that the reliability indicator decreases as the number of tasks increases. This result makes sense, because it is true that the schedulable probability descends if more tasks try to enter the cluster. (b) $\frac{\partial w_p}{\partial \theta} > 0$ means that the indicator rises when the deadline is prolonged.

A factor α is introduced to represent the ratio $\alpha = m/n$, and the distribution of U_m can be developed as

$$U_m \rightarrow \begin{cases} \mathcal{N}(\mu_1, \sigma_1^2) \ \theta \in [0, u_l) \\ \mathcal{N}(\mu_2, \sigma_2^2) \ \theta \in [u_l, u_h) \\ \mathcal{N}(\mu_3, \sigma_3^2) \ \theta \in [u_h, 1] \end{cases}$$

where:

$$\begin{aligned} \mu_1 &= \frac{1}{\theta} \frac{n}{2} \\ \sigma_1 &= \frac{1}{\theta} \sqrt{\frac{n}{12}} \\ \mu_2 &= \frac{\alpha}{\theta} \frac{n}{2} + 2(1 - \alpha)n(1 - \ln 2) \\ \sigma_2 &= \sqrt{\frac{\alpha}{\theta^2} \frac{n}{12} + 4(1 - \alpha)n(\frac{1}{2} - (\ln 2)^2)} \\ \mu_3 &= 2n(1 - \ln 2) \\ \sigma_3 &= \sqrt{4n(\frac{1}{2} - (\ln 2)^2)} \end{aligned} \quad (8.56)$$

The reliability indicators are

$$w_i = \frac{1 - \mu_i}{\sigma_i} \qquad i = 1, 2, 3 \quad (8.57)$$

Gradient vectors are

$$\begin{aligned} \nabla w_1(n, \theta) &= \left(\frac{\partial w_1}{\partial n}, \frac{\partial w_1}{\partial \theta} \right) \\ \nabla w_2(n, \theta, \alpha) &= \left(\frac{\partial w_2}{\partial n}, \frac{\partial w_2}{\partial \theta}, \frac{\partial w_2}{\partial \alpha} \right) \\ \nabla w_3(n) &= \frac{\partial w_3}{\partial n} \end{aligned} \quad (8.58)$$

The reliability indicator of Masrur's DM condition is quite complicated. (a) $\frac{\partial w_i}{\partial n} < 0, i \in [1, 2, 3]$ shows that the accepted ratio of test decreases as the number of tasks increases. (b) $\frac{\partial w_1(n, \theta)}{\partial \theta} > 0$ implies that lengthened deadline can increase the passing probability if the deadline is less than half of the period. (c) The variations of $w_2(n, \theta, \alpha)$ in the θ and α directions are not monotonic any more. Figure 8.7 shows how the value of $w_2(n, \theta, \alpha)$ changes with respect to θ and α. Reliability indicators of the two conditions are both piecewise functions. In order to clearly compare them, a factor is defined as

$$\Delta = w_m - w_p \quad (8.59)$$

The positive value of Δ indicates that a task set is more likely to pass Masrur's test than Peng's test. In other words, Masrur's test is better. Comparison can be detailed by the following four steps.

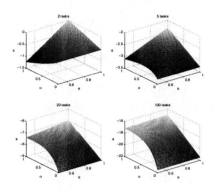

FIGURE 8.7: $w_2(n, \theta, \alpha)$ w.r.t θ and α.

Case $\theta \in [0, 0.5)$:

$$\Delta_1 = w_1 - w_p = 0 \tag{8.60}$$

In this part, the value of Δ is always zero, so the two tests have the same reliability. System designers can choose any of them as the schedulable condition.

Case $\theta \in [0.5, u_l)$:

$$\Delta_2 = w_1 - w_p \tag{8.61}$$

Considering $n \cdot \min u_i \leq \sum_{i=1}^{n} u_i \leq \theta$, another condition $\theta < u_l = (1 + \min u_i)/2 < (1 + \frac{\theta}{n})/2$ is obtained. Therefore, the value of θ falls into range $[0.5, n/(2n - 1))$. Figure 8.8 presents the cases in which Masrur's condition is less

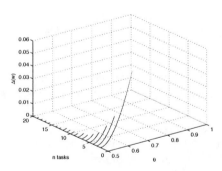

FIGURE 8.8: Better performance of Masrur's test ($\theta \in [0.5, u_l)$).

pessimistic than Peng's. When the deadline is relatively short, Masrur's test performs better. However, this superiority diminishes as more tasks are admitted in the system. That is caused by the possible field $[0.5, n/2n - 1)$ shrinking with the increasing number of tasks.

Case $\theta \in [u_l, u_h)$:

$$\Delta_3 = w_2 - w_p \tag{8.62}$$

Figure 8.9 shows the performance comparison if θ locates in the field $[u_l, u_h)$. The points on each sub-figure stand for the cases where Masrur's condition exceeds Peng's. Especially, Masrur's test is more reliable for most cases when there are only two tasks in the set. Exceeding the number of tasks results in the degradation of Masrur's advantage. The reliability indicator is not only useful for performance

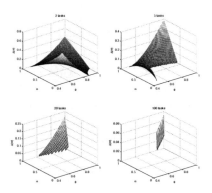

FIGURE 8.9: Better performance of Masrur's test ($\theta \in [u_l, u_h)$).

comparison, but also capable of specifying an exact pattern where the winner can be applied. For example, in Figure 8.9, the system designer can choose a dominant condition based on foreseeable n, α and β. If the point appears on the figure, Masrur's condition wins, otherwise, Peng's test is preferred.

Case: $\theta \in [u_h, 1]$:

$$\Delta_4 = w_3 - w_p \tag{8.63}$$

In this part, one condition needs to be satisfied, that is, $\theta > u_h = (1 + \max u_i)/2 > (1 + \frac{\theta}{n})/2$. The possible field of θ is $[n/(2n - 1), 1]$. In Figure 8.10, only two short lines appear, which stand for the cases where Masrur's test performs better. Clearly, it seldom works as the dominated condition for schedulability test, only under strict constraint that the number of tasks is no more than three. Hence Masrur's condition is not recommended to system designers.

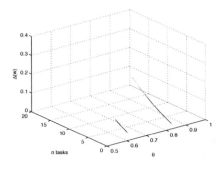

FIGURE 8.10: Better performance of Masrur's test ($\theta \in [u_h, 1]$).

8.7   Concluding remarks

In this chapter, we studied the problem of scheduling real-time tasks on MapReduce cluster, arising from demand for cloud computing. We first formulate the real-time scheduling problem, based on which classic utilization bounds for schedulability test are revisited. We then present a MapReduce scheduling algorithm, combining the particular characteristics of MapReduce. After Map is finished, a proper pause before submission of Reduce can enhance scheduling efficiency for the whole cluster. We deduce the relationship between cluster utilization bound and the ratio of Map to Reduce. This new schedulable bound with segmentation uplifts Liu's bound. The latter can be further considered as a special case of the former. The effectiveness of this bound is evaluated by simulation using SimMapReduce. Results show that this new bound is less pessimistic, and it supports on-line schedulability test in $O(1)$ time complexity. Given the lack of general solutions to compare the performances of different tests, we propose a method to indicate test reliability. Through a reliability indicator, the probability of passing different tests can be compared. We apply this method in several classic schedulability tests. Results show that Liu's salient bound is a dominated condition in RM tests. For DM tests, test reliability depends on system parameters. If these parameters are known in advance, system designers can analyze the performance exactly, and then choose an applicable test among several alternatives.

Chapter 9

Future for cloud computing

Cloud computing implies that computing is not only operated on local computers, but on centralized facilities by third-party computing and storage utilities. It refers to both the applications delivered as services over the Internet and system hardware/-software in datacenters as service providers. Cloud solutions seem to state master keys for the IT enterprises which suffer from budget concerns and economic woes, and a number of industry projects have been started to create a global, multi-data center, open source cloud computing testbed for industry, research, and education.

Encouraging opportunities also brings out corresponding challenges. Cloud computing is easily confused with several existing technologies including grid computing, utility computing, web service, and virtualization. Again, cloud computing is a newly evolved delivery model. It covers with equal importance both technology and business requirements, and it lets users focus on their abilities on demand by abstracting its technology layer. In that case, the scheduling problem in cloud computing is worth reconsidering by researchers and engineers. In this book, we addressed the resource allocation problem in terms of economic aspects to meet business requirements. At the same time, we were concerned with the real-time schedulability test to provide the cloud datacenter with technical supports. Both theoretical and practical efforts were made to solve cloud scheduling problems and to facilitate the succeeding researches.

In order to utilize cloud computing to serve as the infrastructure of multi-dimensional data analysis applications, the combination of traditional parallel database optimization mechanisms and cloud computing is expected. In this book, we try to utilize methods of cloud computing to satisfy commercial software requirements. In addition to realizing a concrete multi-dimensional data analysis query with MapReduce, we mostly focus on the performance optimization, and combine MapReduce with several optimization techniques coming from parallel database. It is an important aspect in designing cloud computing-based solution for business software. This approach does not depend on a third-party product, and it can guarantee the performance application.

Besides the mentioned contributions, our work has raised many interesting questions and issues that deserve further research.

- Choosing suitable serialization/de-serialization algorithms to deal with mapper objects and intermediate results is an important issue. Since these procedures are repeatedly performed, it is closely related to the performance. In our distinct-value-wise job-scheduling, Bitmap compressing is needed to reduce storage requirements and improve the efficiency of data access. We will address these issues in our work in the near future.

- Extending our calculation to a larger computing scale is another interesting direction. In this work, all the experiments were running over one single cluster. However, running experiments over one cluster is a bit far from exploiting a real Cloud platform. In order to further address a more realistic large-scaled multi-dimensional data analytical query processing, multiple clusters or even the real Cloud experimental platform need to be exploited during the experiments. The hardware update will allow us to handle larger datasets.

- Utilizing cloud computing to process large datasets involves another challenging problem—data privacy. This topic was not covered in our work. However, it is still an important aspect. People or enterprises will not want to put their data on the Cloud until their private data can be protected from unauthorized access. Some authorization and authentication technologies have already been developed in Grid Computing. They are very useful for the Cloud platform data accessing protection requirement. However, the authorization and authentication of Cloud platforms are more challenging than in Grid platforms. Since it is a commercialized, shared computing platform, users will require finer authorization and authentication mechanisms.

- Enriching business models for cloud providers. Besides the technical strengths of cloud computing, users decide to head in clouds for economic reasons, so the business model of cloud computing should be more flexible, offering clients scalable price options. For example, Amazon customers can choose purchasing models among on-demand, reserved, spot and even free tier according to their own preferences. With more and more cloud solutions emerging, business models must be reformed to maintain customer loyalty or attract new interest. In addition, new economic models that support the trading, negotiation, provisioning and allocation based on consumer preference should be developed.

- Expanding schedulability bound to more complicated systems. The primer utilization bound for MapReduce cluster is not a final result; our investigation will be continued considering more realistic features of cloud services. We shall extend our result to cases of imprecise computations, dependent tasks, aperiodic tasks, and non-preemptive execution in the future. Since we ideally assume the computation ability of a cluster as a whole by hiding assignment details of every Map/Reduce task inside the cluster, this bound is mainly used in the single processor scenario. Next, we intend to apply this bound and heuristics to solve multi-processor problems.

- Improving reliability of on-line schedulability tests for cloud datacenters. There is always a contradiction between the test accuracy and its time complexity. We have improved the schedulability bound by introducing practical characteristics of MapReduce segmentation, but it is still pessimistic compared with exact schedulability test. Determining test reliability with a low time complexity is still challenging.

There is a lot of other interesting work to do to integrate the MapReduce model into, or utilizing cloud computing in resolving the real problems, including industrial applications as well as scientific computations. For instance, MapReduce's combination with web techniques, re-designing of various algorithms for fitting MapReduce execution style, etc. are all interesting research subjects. We believe that the performance issue addressed in this work represents an important aspect in cloud computing. We hope that our work can provide a useful reference for people who want to study and utilize MapReduce and cloud computing platforms.

Bibliography

[Abadi, 2009] Abadi, D. (2009). Data management in the cloud: limitations and opportunities. *IEEE Data Engineering Bulletin*, 32(1):3–12.

[Abadi et al., 2009] Abadi, D., Boncz, P., and Harizopoulos, S. (2009). Column-oriented database systems. *Proceedings of the VLDB Endowment*, 2(2):1664–1665.

[Abadi et al., 2006] Abadi, D., Madden, S., and Ferreira, M. (2006). Integrating compression and execution in column-oriented database systems. In *SIG-MOD'06: Proceedings of the 2006 ACM SIGMOD International Conference on Management of Data*, pages 671–682, New York, NY, USA. ACM.

[Abadi et al., 2008] Abadi, D., Madden, S., and Hachem, N. (2008). Column-stores versus row-stores: How different are they really? In *SIGMOD'08: Proceedings of the 2008 ACM SIGMOD International Conference on Management of Data*, New York, NY, USA. ACM.

[Abdelzaher and Lu, 2001] Abdelzaher, T. and Lu, C. (2001). Schedulability analysis and utilization bounds for highly scalable real-time service. In *Proceedings of the IEEE Real-Time Technology and Applications Symposium*, pages 15–25.

[Abouzeid et al., 2009] Abouzeid, A., Bajda-Pawlikowski, K., Abadi, D., Silberschatz, A., and Rasin, A. (2009). HadoopDB: an architectural hybrid of MapReduce and DBMS technologies for analytical workloads. In *Proceedings of the VLDB Endowment*, volume 2, pages 922–933.

[Abramson et al., 2002] Abramson, D., Buyya, R., and Giddy, J. (2002). A computational economy for grid computing and its implementation in the Nimrod-G resource broker. *Future Generation Computer Systems*, 18(8):1061–1074.

[Akal et al., 2002] Akal, F., Böhm, K., and Schek, H.-J. (2002). OLAP query evaluation in a database cluster: a performance study on intra-query parallelism. In *ADBIS'02: Proceedings of the 6th East European Conference on Advances in Databases and Information Systems*, pages 218–231, London, UK. Springer-Verlag.

[Akinde et al., 2003] Akinde, M., Bahlen, M., Lakshmanan, L., Johnson, T., and Srivastava, D. (2003). Efficient OLAP query processing in distributed data warehouses. *Information Systems*, 28(1–2):111–135.

[Amazon EC2, 2011] Amazon EC2 (2011). Web Published. Available online at: http://aws.amazon.com/ec2/ (accessed January 1st, 2012).

[Amazon S3, 2011] Amazon S3 (2011). Web Published. Available online at: http://aws.amazon.com/s3/ (accessed January 1st, 2012).

[An et al., 2007] An, B., Miao, C., and Shen, Z. (2007). Market based resource allocation with incomplete information. In *Proceedings of the 20th International Joint Conference on Artificial Intelligence*, pages 1193–1198, San Francisco, CA, USA. Morgan Kaufmann Publishers Inc.

[Andersson et al., 2001] Andersson, B., Baruah, S., and Jonsson, J. (2001). Static-priority scheduling on multi-processors. In *RTSS'01: Proceedings of the 22nd IEEE Real-Time Systems Symposium*, page 93, Washington, DC, USA. IEEE Computer Society.

[Armbrust et al., 2009] Armbrust, M., Fox, A., and Griffith, R. (2009). Above the clouds: a Berkeley view of cloud computing. Technical Report UCB/EECS-2009-28, EECS Department, University of California, Berkeley, CA, USA.

[Aster nCluster, 2012] Aster nCluster (2012). Aster nCluster: in-database MapReduce. Web Published. Available on-line at: http://www.asterdata.com/product/mapreduce.php (accessed January 1st, 2012).

[Audsley et al., 1995] Audsley, N., Burns, A., Davis, R., Tindell, K., and Wellings, A. (1995). Fixed priority pre-emptive scheduling: a historical perspective. *Real-Time Systems*, 8(2-3):173–198.

[Audsley et al., 1993] Audsley, N., Burns, A., Richardson, M., Tindell, K., and Wellings, A. (1993). Applying new scheduling theory to static priority pre-emptive scheduling. *Software Engineering Journal*, 8:284–292.

[Baker, 2005] Baker, T. (2005). An analysis of EDF schedulability on a multi-processor. *IEEE Transactions on Parallel and Distributed Systems*, 16:760–768.

[Baruah and Goossens, 2003] Baruah, S. and Goossens, J. (2003). Rate-monotonic scheduling on uniform multi-processors. *IEEE Transaction on Computers*, 52:966–970.

[BEinGRID, 2011] BEinGRID (2011). Web Published. Available online at: http://www.beingrid.eu/ (accessed January 1st, 2012).

[Bellatreche and Boukhalfa, 2005] Bellatreche, L. and Boukhalfa, K. (2005). An evolutionary approach to Schema partitioning selection in a data warehouse. In *Proceedings of Data Warehousing and Knowledge Discovery*, pages 115–125, Copenhagen, Denmark. Springer.

[Bellatreche et al., 2004] Bellatreche, L., Schneider, M., Lorinquer, H., and Mohania, M. (2004). Bringing together partitioning, materialized views and indexes to optimize performance of relational data warehouses. In *Proceedings of the International Conference on Data Warehousing and Knowledge Discovery*, pages 15–25.

[Bernardino and Madeira, 2012] Bernardino, J. and Madeira, H. (2012). Data warehousing and OLAP: improving query performance using distributed computing. Web Published. Available online at: `http://citeseerx.ist.psu.edu/viewdoc/summary?doi=10.1.1.35.9000` (accessed January 1st, 2012).

[Bini and Baruah, 2007] Bini, E. and Baruah, S. (2007). Efficient computation of response time bounds under fixed-priority scheduling. In *RTNS: Proceedings of the 15th Real-Time and Network Systems*, pages 95–104.

[Bini and Buttazzo, 2002] Bini, E. and Buttazzo, G. (2002). The space of rate monotonic schedulability. In *Proceedings of the IEEE Real-Time Systems Symposium*, pages 169–178.

[Bini and Buttazzo, 2004] Bini, E. and Buttazzo, G. (2004). Schedulability analysis of periodic fixed priority systems. *IEEE Transactions on Computers*, 53(11):1462–1473.

[Bini et al., 2003] Bini, E., Buttazzo, G., and Buttazzo, G. (2003). Rate monotonic analysis: the hyperbolic bound. *IEEE Transactions on Computers*, 52(7):933–942.

[Bitmap, 2012] Bitmap (2012). Web Published. Available on-line at `http://en.wikipedia.org/wiki/Bitmap_index\#Compression` (accessed January 1st, 2012).

[Böhm et al., 2010] Böhm, M., Leimeister, S., Riedl, C., and Krcmar, H. (2010). Cloud computing: Outsourcing 2.0 or a new business model for IT provisioning? In Keuper, F., Oecking, C., and Degenhardt, A., editors, *Application Management*. Gabler, Wiesbaden.

[Boral et al., 1990] Boral, H., Alexander, W., Clay, L., Copeland, G., Danforth, S., Franklin, M., Hart, B., Smith, M., and Valduriez, P. (1990). Prototyping Bubba: a highly parallel database system. *IEEE Transactions on Knowledge and Data Engineering*, 2:4–24.

[Borthakur, 2007] Borthakur, D. (2007). *The Hadoop distributed file system: architecture and design*. The Apache Software Foundation.

[Braun et al., 2001] Braun, T., Siegel, H.-J., Beck, N., Bölöni, L., Maheswaran, M., Reuther, A., Robertson, J., Theys, M., Yao, B., Hensgen, D., and Freund, R. (2001). A comparison of eleven static heuristics for mapping a class of independent tasks onto heterogeneous distributed computing systems. *Journal of Parallel and Distributed Computing*, 61:810–837.

[Bredin et al., 2003] Bredin, J., Kotz, D., Rus, D., Maheswaran, R., Imer, C., and Basar, T. (2003). Computational markets to regulate mobile-agent systems. *Autonomous Agents and Multi-Agent Systems*, 6:235–263.

[Brewer, 2005] Brewer, E. (2005). *Readings in database systems*, chapter Combining systems and databases: a search engine retrospective. MIT Press, Cambridge, MA, USA, 4th edition.

[Bril et al., 2003] Bril, R., Verhaegh, W., and Pol, E.-J. (2003). Initial values for online response time calculations. In *Proceedings of the 15th Euromicro Conference on Real-Time Systems*, pages 13–22.

[Burchard et al., 1994] Burchard, A., Liebeherr, J., Oh, Y., and Son, S. (1994). Assigning real-time tasks to homogeneous multi-processor systems. Technical report, University of Virginia, Charlottesville, VA, USA.

[Buyya et al., 2002] Buyya, R., Abramson, D., Giddy, J., and Stockinger, H. (2002). Economic models for resource management and scheduling in grid computing. In *Concurrency and Computation: Practice and Experience*, pages 1507–1542. Wiley Press.

[Buyya et al., 2009a] Buyya, R., Pandey, S., and Vecchiola, C. (2009a). Cloudbus toolkit for market-oriented cloud computing. In *CloudCom'09: Proceedings of the 1st International Conference on Cloud Computing*, pages 24–44, Berlin, Heidelberg. Springer-Verlag.

[Buyya et al., 2009b] Buyya, R., Ranjan, R., and Calheiros, R. (2009b). Modeling and simulation of scalable cloud computing environments and the CloudSim toolkit: challenges and opportunities. In *Proceedings of the 7th High Performance Computing and Simulation Conference*, Leipzig, Germany. IEEE Press.

[Buyya et al., 2009c] Buyya, R., Yeoa, C.-S., Venugopala, S., Broberg, J., and Brandic, I. (2009c). Cloud computing and emerging IT platforms: vision, hype, and reality for delivering computing as the 5th utility. *Future Generation Computer Systems*, 25(6):599–616.

[Cafaro and Aloisio, 2010] Cafaro, M. and Aloisio, G. (2010). *Grids, clouds and virtualization*. Springer Publishing Company, Incorporated, 1st edition.

[Chang et al., 2008] Chang, F., Dean, J., Ghemawat, S., Hsieh, W., Wallach, D., Burrows, M., Chandra, T., Fikes, A., and Gruber, R. (2008). Bigtable: a distributed storage system for structured data. *ACM Transaction on Computer Systems*, 26(2):1–26.

[Chaudhuri and Dayal, 1997] Chaudhuri, S. and Dayal, U. (1997). An overview of data warehousing and OLAP technology. *SIGMOD Rec.*, 26(1):65–74.

[Chen et al., 2003] Chen, D., Mok, A., and Kuo, T.-W. (2003). Utilization bound revisited. *IEEE Transactions on Computers*, 52:351–361.

[Chen et al., 2008] Chen, L., Olston, C., and Ramakrishnan, R. (2008). Parallel evaluation of composite aggregate queries. In *Proceedings of the International Conference on Data Engineering*, pages 218–227, Los Alamitos, CA, USA. IEEE Computer Society.

[Christopher et al., 2008] Christopher, O., Benjamin, R., Utkarsh, S., Ravi, K., and Andrew, T. (2008). Pig Latin: a not-so-foreign language for data processing. In *SIGMOD'08: Proceedings of the 2008 ACM SIGMOD International Conference on Management of Data*, pages 1099–1110, New York, NY, USA. ACM.

[Cluster-On-Demand, 2011] Cluster-On-Demand (2011). Web Published. Available online at: http://www.cs.duke.edu/nicl/cod/ (accessed January 1st, 2012).

[Cutting and Pedersen, 1990] Cutting, D. and Pedersen, J. (1990). Optimization for dynamic inverted index maintenance. In *SIGIR'90: Proceedings of the 13th Annual International ACM SIGIR Conference on Research and Development in Information Retrieval*, pages 405–411, New York, NY, USA. ACM.

[D'Agents, 2011] D'Agents (2011). Web Published. Available online at: http://agent.cs.dartmouth.edu/ (accessed January 1st, 2012).

[Dailianas et al., 2000] Dailianas, A., Yemini, Y., Florissi, D., and Huang, H. (2000). MarketNet: market-based protection of network systems and services - an application to SNMP protection. In *Proceedings of 19th IEEE International Conference on Computer Communications*, pages 1391–1400.

[DataGrid Project, 2012] DataGrid Project (2012). Web Published. Available online at: http://eu-datagrid.web.cern.ch/eu-datagrid/ (accessed January 1st, 2012).

[Datta et al., 1998] Datta, A., Moon, B., and Thomas, H. (1998). A case for parallelism in data warehousing and OLAP. In *DEXA'98: Proceedings of the 9th International Workshop on Database and Expert Systems Applications*, page 226, Washington, DC, USA. IEEE Computer Society.

[Datta et al., 2012] Datta, A., Vandermeer, D., Ramamritham, K., and Moon, B. (2012). Applying parallel processing techniques in data warehousing and OLAP. Web Published. Available online at: http://citeseerx.ist.psu.edu/viewdoc/summary?doi=10.1.1.32.5528 (accessed January 1st, 2012).

[Dean and Ghemawat, 2004] Dean, J. and Ghemawat, S. (2004). MapReduce: simplified data processing on large clusters. In *OSID'04: Proceedings of the International Conference on Operating Systems Design and Implementation*, pages 137–150.

[Dean and Ghemawat, 2008] Dean, J. and Ghemawat, S. (2008). MapReduce: simplified data processing on large clusters. *Communications of the ACM*, 51(1):107–113.

[Deshpande et al., 1998] Deshpande, P., Ramasamy, K., Shukla, A., and Naughton, J. (1998). Caching multi-dimensional queries using chunks. In *SIGMOD Rec.*, volume 27, pages 259–270, New York, NY, USA. ACM.

[DeWitt et al., 1986] DeWitt, D., Gerber, R., Graefe, G., Heytens, M., Kumar, K., and Muralikrishna, M. (1986). GAMMA: a high performance dataflow database machine. In *VLDB'86: Proceedings of the 12th International Conference on Very Large Data Bases*, pages 228–237. Morgan Kaufmann.

[DeWitt and Gray, 1992] DeWitt, D. and Gray, J. (1992). Parallel database systems: the future of high performance database systems. *Communication ACM*, 35(6):85–98.

[DeWitt and Stonebraker, 2012] DeWitt, D. and Stonebraker, M. (2012). MapReduce: a major step backwards. Web Published. Available online at: `http://databasecolumn.vertica.com/database-innovation/mapreduce-a-major-step-backwards/` (accessed January 1st, 2012).

[Dhall and Liu, 1978] Dhall, S.-K. and Liu, C.-L. (1978). On a real-time scheduling problem. *Operations Research*, 26(1):127–140.

[DIET, 2011] DIET (2011). Web Published. Available online at: `http://graal.ens-lyon.fr/DIET/` (accessed January 1st, 2012).

[Douglas, 1979] Douglas, C. (1979). Ubiquitous B-tree. *ACM Computing Surveys*, 11(2):121–137.

[Dryad, 2011] Dryad (2011). Web Published. Available online at: `http://research.microsoft.com/en-us/projects/dryad/` (accessed January 1st, 2012).

[Elmroth and Tordsson, 2005] Elmroth, E. and Tordsson, J. (2005). A grid resource broker supporting advance reservations and benchmark-based resource selection. In *Lecture Notes in Computer Science*, pages 1061–1070. Springer-Verlag.

[Fan et al., 2009] Fan, Q., Wu, Q., Magoulès, F., Xiong, N., Vasilakos, A., and He, Y. (2009). Game and balance multicast architecture algorithms for sensor grid. *Sensors*, 9(9):7177–7202.

[Fisher and Baruah, 2006] Fisher, N. and Baruah, S. (2006). A fully polynomial-time approximation scheme for feasibility analysis in static-priority systems with bounded relative deadlines. *Journal of Embedded Computing*, 2(3-4):291–299.

[Foster et al., 2001] Foster, I., Kesselman, C., and Tuecke, S. (2001). The anatomy of the grid: enabling scalable virtual organizations. *International Journal of Supercomputer Applications*, 15(3):200–222.

[Foster et al., 2008] Foster, I., Zhao, Y., Raicu, I., and Lu, S. (2008). Cloud computing and grid computing 360-degree compared. In *GCE'08: Proceedings of the Grid Computing Environments Workshop, 2008*, pages 1–10.

[Fushimi et al., 1986] Fushimi, S., Kitsuregawa, M., and Tanaka, H. (1986). An overview of the system software of a parallel relational database machine GRACE. In *VLDB'86: Proceedings of the 12th International Conference on Very Large Data Bases*, pages 209–219, San Francisco, CA, USA. Morgan Kaufmann Publishers Inc.

[FutureGrid, 2011] FutureGrid (2011). Web Published. Available online at: `https://portal.futuregrid.org/` (accessed January 1st, 2012).

[Galstyan et al., 2003] Galstyan, A., Kolar, S., and Lerman, K. (2003). Resource allocation games with changing resource capacities. In *AAMAS 2003: Proceedings of the International Conference on Autonomous Agents and Multi-Agent Systems*, pages 145–152. ACM Press.

[Ghemawat et al., 2003] Ghemawat, S., Gobioff, H., and Leung, S.-T. (2003). The Google file system. *Operating Systems Review*, 37(5):29–43.

[Gibbons, 1992] Gibbons, R. (1992). *A primer in game theory*. Pearson Higher Education.

[Goil and Choudhary, 1999] Goil, S. and Choudhary, A. (1999). A parallel scalable infrastructure for OLAP and data mining. In *IDEAS'99: Proceedings of the 1999 International Symposium on Database Engineering & Applications*, page 178, Washington, DC, USA. IEEE Computer Society.

[Google App Engine, 2011] Google App Engine (2011). Web Published. Available online at: `http://code.google.com/appengine/` (accessed January 1st, 2012).

[Graefe, 1990] Graefe, G. (1990). Encapsulation of parallelism in the Volcano query processing system. In *SIGMOD'90: Proceedings of the 1990 ACM SIGMOD International Conference on Management of Data*, pages 102–111, New York, NY, USA. ACM.

[Graefe, 1993] Graefe, G. (1993). Query evaluation techniques for large databases. *ACM Computing Surveys*, 25:73–170.

[Greenplum, 2012] Greenplum (2012). Web Published. Available on-line at: `http://www.greenplum.com/resources/MapReduce/` (accessed January 1st, 2012).

[Grid'5000, 2012] Grid'5000 (2012). Available on-line at: `https://www.grid5000.fr/` (accessed January 1st, 2012).

[GridGain, 2012] GridGain (2012). Web Published. Available on-line at: `http://www.gridgain.com/` (accessed January 1st, 2012).

[Hadoop, 2012a] Hadoop (2012a). Web Published. Available on-line at: `http://hadoop.apache.org/` (accessed January 1st, 2012).

[Hadoop, 2012b] Hadoop (2012b). Hadoop distributed file system. Web Published. Available on-line at: `http://hadoop.apache.org/hdfs/` (accessed January 1st, 2012).

[Han, 1998] Han, C.-C. (1998). A better polynomial-time schedulability test for real-time multi-frame tasks. In *Proceedings of the IEEE Real-Time Technology and Applications Symposium*, pages 104–113.

[Harter and Paul, 1987] Harter, J. and Paul, K. (1987). Response times in level-structured systems. *ACM Transaction on Computer Systems*, 5:232–248.

[Hasan, 1996] Hasan, W. (1996). *Optimization of SQL queries for parallel machines*. PhD thesis, Stanford, CA, USA.

[Hellerstein, 2012] Hellerstein, J. (2012). Parallel programming in the age of big data. Web Published. Available on-line at: `http://gigaom.com/2008/11/09/mapreduce-leads-the-way-for-parallel-programming/` (accessed January 1st, 2012).

[Hive, 2012] Hive (2012). Web Published. Available on-line at: `http://hadoop.apache.org/hive/` (accessed January 1st, 2012).

[Hyman et al., 1991] Hyman, J., Lazar, A., and Pacifici, G. (1991). Real-time scheduling with quality of service constraints. *IEEE Journal on Selected Areas in Communications*, 9:1052–1063.

[Isard et al., 2009] Isard, M., Prabhakaran, V., Currey, J., Wieder, U., Talwar, K., and Goldberg, A. (2009). Quincy: fair scheduling for distributed computing clusters. In *SOSP'09: Proceedings of the ACM SIGOPS 22nd Symposium on Operating Systems Principles*, pages 261–276, New York, NY, USA. ACM.

[Jim et al., 1998] Jim, G., Surajit, C., Adam, B., Andrew, L., Don, R., Murali, V., Frank, P., and Hamid, P. (1998). *Data cube: a relational aggregation operator generalizing group-by, cross-tab, and sub-totals*, pages 555–567. Morgan Kaufmann Publishers, 3rd edition.

[Joseph and Pandya, 1986] Joseph, M. and Pandya, P. (1986). Finding response times in a real-time system. *The Computer Journal*, 29:390–395.

[Kale et al., 2004] Kale, L., Kumar, S., Potnuru, M., DeSouza, J., and Bandhakavi, S. (2004). Faucets: efficient resource allocation on the computational grid. In *ICPP'04: Proceedings of the 2004 International Conference on Parallel Processing*, pages 396–405, Washington, DC, USA. IEEE Computer Society.

[Kato et al., 2009] Kato, S., Rajkumar, R., and Ishikawa, Y. (2009). A loadable real-time scheduler suite for multi-core platform. Technical Report 12, Carnegie Mellon University, Department of Electrical and Computer Engineering.

[Keith and Burkhard, 2010] Keith, J. and Burkhard, N.-L., editors (2010). *The future of cloud computing: opportunities for European cloud computing beyond 2010*. Available online at: `http://cordis.europa.eu/fp7/ict/ssai/docs/cloud-report-final.pdf` (accessed January 1st, 2012).

[Kephart and Chess, 2003] Kephart, J. and Chess, D. (2003). The vision of autonomic computing. *Computer*, 36:41–50.

[Khan and Ahmad, 2006] Khan, S.-U. and Ahmad, I. (2006). Non-cooperative, semi-cooperative, and cooperative games-based grid resource allocation. In *Proceedings of the International Parallel and Distributed Processing Symposium*, volume 0, page 101, Los Alamitos, CA, USA. IEEE Computer Society.

[Kossmann, 2000] Kossmann, D. (2000). The state-of-the-art in distributed query processing. *ACM Computing Surveys*, 32(4):422–469.

[Kotowski et al., 2007] Kotowski, N., Lima, A., Pacitti, E., Valduriez, P., and Mattoso, M. (2007). OLAP query processing in grids. *Concurrency and Computation: Practice and Experience*, 20(17):2039–2048.

[Kuo and Mok, 1991] Kuo, T.-W. and Mok, A.-K. (1991). Load adjustment in adaptive real-time systems. In *Proceedings of the IEEE Real-Time Technology and Applications Symposium*, pages 160–171.

[Kwok et al., 2005] Kwok, Y.-K., Song, S., and Hwang, K. (2005). Selfish grid computing: game-theoretic modeling and Nash performance results. In *Proceedings of the International Symposium on Cluster Computing and the Grid*, pages 9–12.

[Ladjel et al., 1999] Ladjel, B., Kamalakar, K., and Mukesh, M. (1999). OLAP query processing for partitioned data warehouses. In *DANTE'99: Proceedings of the 1999 International Symposium on Database Applications in Non-Traditional Environments*, Washington, DC, USA. IEEE Computer Society.

[Lämmel, 2007] Lämmel, R. (2007). Google's MapReduce programming model: revisited. *Science of Computer Programming*, 68(3):208–237.

[Lehoczky, 1990] Lehoczky, J. (1990). Fixed priority scheduling of periodic task sets with arbitrary deadlines. In *Proceedings of the 11th Real-Time Systems Symposium*, pages 201–209.

[Lehoczky et al., 1989] Lehoczky, J., Sha, L., and Ding, Y. (1989). The rate monotonic scheduling algorithm: exact characterization and average case behavior. In *Proceedings of the IEEE Real-Time Technology and Applications Symposium*, pages 166–171.

[Leung and Whitehead, 1982] Leung, J.-Y.-T. and Whitehead, J. (1982). On the complexity of fixed-priority scheduling of periodic, real-time tasks. *Performance Evaluation*, 2:237–250.

[Li et al., 2004] Li, X., Han, J., and Gonzalez, H. (2004). High-dimensional OLAP: a minimal cubing approach. In *Proceedings of the 30th VLDB Conference*, Toronto, Canada.

[Liao and Pei, 2008] Liao, H.-M. and Pei, G.-S. (2008). Cache-based aggregate query shipping: an efficient scheme of distributed OLAP query processing. *Journal of Computer Science and Technology*, 23(6):905–915.

[Lima et al., 2004a] Lima, A., Mattoso, M., and Valduriez, P. (2004a). Adaptive virtual partitioning for OLAP query processing in a database cluster. In *SBBD 2004: Proceedings of the 19th Brazilian Symposium on Database Systems, Brasilia, Brazil, October 18-20*, pages 92–105.

[Lima et al., 2004b] Lima, A., Mattoso, M., and Valduriez, P. (2004b). OLAP query processing in a database cluster. In *Proceedings of 10th International Euro-Par Conference*, pages 355–362, Pisa, Italy. Springer.

[Lin et al., 2009] Lin, J., Konda, S., and Mahindrakar, S. (2009). Low-latency, high-throughput access to static global resources within the Hadoop framework. Web Published. Available on-line at: http://www.umiacs.umd.edu/~jimmylin/publications/Lin_etal_TR2009.pdf (accessed January 1st, 2012).

[Liu and Layland, 1973] Liu, C.-L. and Layland, J. (1973). Scheduling algorithms for multi-programming in a hard-real-time environment. *Journal of the Association for Computing Machinery*, 20(1):46–61.

[López et al., 2004] López, J.-M., Díaz, J.-L., and García, D.-F. (2004). Minimum and maximum utilization bounds for multi-processor rate monotonic scheduling. *IEEE Transactions Parallel Distributed Systems*, 15(7):642–653.

[Lu et al., 2006] Lu, W.-C., Hsieh, J.-W., and Shih, W.-K. (2006). A precise schedulability test algorithm for scheduling periodic tasks in real-time systems. In *SAC'06: Proceedings of the 2006 ACM Symposium on Applied Computing*, pages 1451–1455, New York, NY, USA. ACM.

[Magoulès, 2009] Magoulès, F., editor (2009). *Fundamentals of Grid Computing: Theory, Algorithms and Technologies*. Chapman & Hall/CRC Numerical Analysis & Scientific Computing. CRC Press, Boca Raton, FL, USA.

[Magoulès et al., 2008] Magoulès, F., Nguyen, T.-H.-M., and Yu, L. (2008). *Grid Resource Management: Towards Virtual and Services Compliant Grid Computing*. Chapman & Hall/CRC Numerical Analysis & Scientific Computing. CRC Press, Boca Raton, FL, USA.

[Magoulès et al., 2009] Magoulès, F., Pan, J., Tan, K.-A., and Kumar, A. (2009). *Introduction to Grid Computing*. Chapman & Hall/CRC Numerical Analysis & Scientific Computing. CRC Press, Boca Raton, FL, USA.

[Maheswaran and Basar, 2003] Maheswaran, R. and Basar, T. (2003). Nash equilibrium and decentralized negotiation in auctioning divisible resources. *Group Decision and Negotiation*, 12:361–395.

[Manabe and Aoyagi, 1995] Manabe, Y. and Aoyagi, S. (1995). A feasibility decision algorithm for rate monotonic scheduling of periodic real-time tasks. In *Proceedings of the IEEE Real-Time Technology and Applications Symposium*, pages 212–218.

[Masrur and Chakraborty, 2011] Masrur, A. and Chakraborty, S. (2011). Near-optimal constant-time admission control for DM tasks via non-uniform approximations. In *RTAS: Proceedings of the 17th IEEE Real-Time and Embedded Technology and Applications Symposium*, Chicago, IL, USA.

[Masrur et al., 2010] Masrur, A., Chakraborty, S., and Färber, G. (2010). Constant-time admission control for deadline monotonic tasks. In *DATE: Proceedings of the Conference on Design, Automation and Test in Europe*, pages 220–225.

[Maui, 2011] Maui (2011). Maui cluster scheduler. Web Published. Available online at: `http://www.clusterresources.com/products/maui-cluster-scheduler.php` (accessed January 1st, 2012).

[Mell and Grance, 2010] Mell, P. and Grance, T. (2010). The NIST definition of cloud computing. *National Institute of Standards and Technology*, 53:7.

[Mosix, 2011] Mosix (2011). Web Published. Available online at: `http://www.mosix.cs.huji.ac.il/` (accessed January 1st, 2012).

[Oh and Bakker, 1998] Oh, D.-I. and Bakker, T.-P. (1998). Utilization bounds for n-processor rate monotone scheduling with static processor assignment. *Real-Time Systems*, 15(2):183–192.

[O'Neil and Quass, 1997] O'Neil, P. and Quass, D. (1997). Improved query performance with variant indexes. In *Proceedings of the 1997 ACM SIGMOD International Conference on Management of Data*, pages 38–49. ACM.

[OpenNebula, 2011] OpenNebula (2011). Web Published. Available online at: `http://dev.opennebula.org/` (accessed January 1st, 2012).

[Oracle, 2011] Oracle (2011). Oracle Grid Engine. Web Published. Available online at: `http://www.sun.com/software/sge/` (accessed January 1st, 2012).

[Pan et al., 2010a] Pan, J., Biannic, Y. L., and Magoulès, F. (2010a). Parallelizing multiple group-by query in share-nothing environment: a MapReduce study case. In *Proceedings of the 19th ACM International Symposium on High Performance Distributed Computing (HPDC)*, pages 856–863. ACM Press.

[Pan et al., 2010b] Pan, J., Magoulès, F., and Biannic, Y. L. (2010b). Executing multiple group-by query in a mapreduce approach. In *International Conference on Digital Business (ICDB) and Second International Conference on Communication Systems, Networks and Applications (ICCSNA), Hong-Kong, June 29–July 1, 2010*, volume 2, pages 38–41. IEEE Computer Society.

[Pan et al., 2010c] Pan, J., Magoulès, F., and Biannic, Y. L. (2010c). Executing multiple group-by query using MapReduce approach: implementation and optimization. In Bellavista, P., Chang, R.-S., Lin, S., and Sloot, P., editors, *Advances in Grid and Pervasive Computing. Proceedings of the 5th International Conference in Grid and Pervasive Computing (GPC 2010), Hualien, Taiwan, May 10–13, 2010*, volume 6104 of *Lecture Notes in Computer Science (LNCS)*, pages 652–661. Springer-Verlag.

[Pan et al., 2010d] Pan, J., Magoulès, F., and Biannic, Y. L. (2010d). Implementing and optimizing multiple group-by query in a MapReduce approach. *Journal of Algorithms and Computational Technology*, 4(2):183–206.

[Parashar and Hariri, 2005] Parashar, M. and Hariri, S. (2005). Autonomic computing: an overview. In *Unconventional Programming Paradigms*, pages 247–259. Springer Verlag.

[Park et al., 1995] Park, D.-W., Natarajan, S., Kanevsky, A., and Kim, M. (1995). A generalized utilization bound test for fixed-priority real-time scheduling. In *RTCSA'95: Proceedings of the 2nd International Workshop on Real-Time Computing Systems and Applications*, pages 73–77, Washington, DC, USA. IEEE Computer Society.

[Passing, 2012] Passing, J. (2012). The Google file system and its application in MapReduce. Web Published. Available on-line at: `http://int3.de/res/GfsMapReduce/GfsMapReducePaper.pdf` (accessed January 1st, 2012).

[Pavlo et al., 2009] Pavlo, A., Paulson, E., Rasin, A., Abadi, D., DeWitt, D., Madden, S., and Stonebraker, M. (2009). A comparison of approaches to large-scale data analysis. In *SIGMOD'09: Proceedings of the 35th SIGMOD International Conference on Management of Data*, pages 165–178, New York, NY, USA. ACM.

[Peng and Shin, 1993] Peng, D.-T. and Shin, K.-G. (1993). A new performance measure for scheduling independent real-time tasks. *Journal of Parallel Distributed Computing*, 19:11–26.

[Plummer et al., 2008] Plummer, D., Cearley, D., and Smith, D. (2008). Cloud computing confusion leads to opportunity. Technical Report G00159034, Gartner Research.

[Ragusa et al., 2009] Ragusa, C., Longo, F., and Puliafito, A. (2009). Experiencing with the cloud over gLite. In *CLOUD'09: Proceedings of the 2009 ICSE Workshop on Software Engineering Challenges of Cloud Computing*, pages 53–60, Washington, DC, USA. IEEE Computer Society.

[Ronnie et al., 2008] Ronnie, C., Bob, J., Perake, L., Bill, R., Darren, S., Simon, W., and Jingren, Z. (2008). SCOPE: Easy and efficient parallel processing of massive data sets. volume 1, pages 1265–1276.

[Roure et al., 2004] Roure, D. D., Baker, M., Jennings, N., and Shadbolt, N. (2004). The evolution of the grid. In *Grid Computing: Making the Global Infrastructure a Reality*, pages 65–100. John Wiley & Sons.

[Run-length-encoding, 2012] Run-length-encoding (2012). Web Published. Available on-line at `http://en.wikipedia.org/wiki/Run-length_encoding` (accessed January 1st, 2012).

[Sanjay and Alok, 1997] Sanjay, G. and Alok, C. (1997). High performance OLAP and data mining on parallel computers. In *Proceedings of the International Conference on Data Mining Knowledge Discovery*, volume 1, pages 391–417, Hingham, MA, USA. Kluwer Academic Publishers.

[Sha et al., 2004] Sha, L., Abdelzaher, T., Arzen, K.-E., Cervin, A., Baker, T., Burns, A., Buttazzo, G., Caccamo, M., Lehoczky, J., and Mok, A. (2004). Real-time scheduling theory: a historical perspective. *Real-Time Systems*, 28:101–155.

[Sha and Goodenough, 1990] Sha, L. and Goodenough, J. (1990). Real-time scheduling theory and Ada. *Computer*, 23(4):53–62.

[Shatdal and Naughton, 1995] Shatdal, A. and Naughton, J. (1995). Adaptive parallel aggregation algorithms. In *SIGMOD'95: Proceedings of the 1995 ACM SIGMOD International Conference on Management of Data*, New York, NY, USA. ACM.

[Shoukat et al., 1999] Shoukat, M. M., Maheswaran, M., Ali, S., Siegel, H.-J., Hensgen, D., and Freund, R. (1999). Dynamic mapping of a class of independent tasks onto heterogeneous computing systems. *Journal of Parallel and Distributed Computing*, 59:107–131.

[Sismanis et al., 2002] Sismanis, Y., Deligiannakis, A., Roussopoulos, N., and Kotidis, Y. (2002). Dwarf: shrinking the PetaCube. In *SIGMOD'02: Proceedings of the 2002 ACM SIGMOD International Conference on Management of Data*, pages 464–475, New York, NY, USA. ACM.

[Sjödin and Hansson, 1998] Sjödin, M. and Hansson, H. (1998). Improved response-time analysis calculations. In *Proceedings of the IEEE Real-Time Technology and Applications Symposium*, pages 399–408.

[SLA@SOI, 2011] SLA@SOI (2011). Web Published. Available online at: `http://sla-at-soi.eu/` (accessed January 1st, 2012).

[Stamos and Young, 1993] Stamos, J. and Young, H. (1993). A symmetric fragment and replicate algorithm for distributed joins. *IEEE Transactions on Parallel Distributed Systems*, 4(12):1345–1354.

[Stanford University, 2011] Stanford University (2011). Stanford Peers. Web Published. Available online at: http://infolab.stanford.edu/peers/ (accessed January 1st, 2012).

[Stephano et al., 1982] Stephano, C., Mauro, N., and Pelagatti, G. (1982). Horizontal data partitioning in database design. In *SIGMOD'82: Proceedings of the 1982 ACM SIGMOD International Conference on Management of Data*, pages 128–136. ACM.

[Stockinger et al., 2002] Stockinger, K., Wu, K., and Shoshani, A. (2002). Strategies for processing ad-hoc queries on large data warehouses. In *DOLAP'02: Proceedings of the 5th ACM International Workshop on Data Warehousing and OLAP*, pages 72–79, New York, NY, USA. ACM.

[Sun et al., 2009] Sun, Y., Tilak, S., Thulasiram, R., and Chiu, K. (2009). *Markets, mechanisms, games, and their implications in grids*, chapter 2, pages 29–48. John Wiley & Sons, Inc.

[Teng and Magoulès, 2010] Teng, F. and Magoulès, F. (2010). A new game theoretical ressource allocation algorithm for cloud computing. In Bellavista, P., Chang, R.-S., and Sloot, S. L. P., editors, *Advances in Grid and Pervasive Computing*, volume 6104 of *Lecture Notes in Computer Science (LNCS)*, pages 321–330. Springer.

[Thain et al., 2005] Thain, D., Tannenbaum, T., and Livny, M. (2005). Distributed computing in practice: the Condor experience. *Concurrency and Computation: Practice and Experience*, 17:323–356.

[Thuel and Lehoczky, 1994] Thuel, S. and Lehoczky, J. (1994). Algorithms for scheduling hard aperiodic tasks in fixed-priority systems using slack stealing. In *Proceedings of the IEEE Real-Time Technology and Applications Symposium*, pages 22–33.

[Tokuda and Mercer, 1989] Tokuda, H. and Mercer, C.-W. (1989). ARTS: a distributed real-time kernel. *Operating Systems Review*, 23:29–53.

[Ucar et al., 2006] Ucar, B., Aykanat, C., Kaya, K., and Ikinci, M. (2006). Task assignment in heterogeneous computing systems. *Journal of Parallel and Distributed Computing*, 66(1):32–46.

[Uthaisombut, 2008] Uthaisombut, P. (2008). Generalization of EDF and LLF: identifying all optimal online algorithms for minimizing maximum lateness. *Algorithmica*, 50:312–328.

[Valduriez, 1987] Valduriez, P. (1987). Join indices. *ACM Transactions on Database Systems*, 12(2):218–246.

[Venugopal et al., 2009] Venugopal, S., Broberg, J., and Buyya, R. (2009). Open-PEX: an open provisioning and execution system for virtual machines. Technical Report CLOUDS-TR-2009-8, CLOUDS Laboratory, The University of Melbourne, Australia,.

[Wang et al., 2002] Wang, W., Lu, H., Feng, J., and Xu-Yu, J. (2002). Condensed Cube: an efficient approach to reducing data cube size. In *Proceedings of the International Conference on Data Engineering*, page 0155, Los Alamitos, CA, USA. IEEE Computer Society.

[Wei et al., 2009] Wei, G., Vasilakos, A., Zheng, Y., and Xiong, N. (2009). A game-theoretic method of fair resource allocation for cloud computing services. *Journal of Supercomputing*, 54:1–18.

[Xtreemos, 2011] Xtreemos (2011). Web Published. Available online at: http://www.xtreemos.eu/ (accessed January 1st, 2012).

[Yang et al., 2007] Yang, H.-C., Dasdan, A., Hsiao, R.-L., and Parker, S. (2007). Map-reduce-merge: simplified relational data processing on large clusters. In *SIGMOD'07: Proceedings of the 2007 ACM SIGMOD International Conference on Management of Data*, pages 1029–1040. ACM.

[Yi et al., 2010] Yi, S., Kondo, D., and Andrzejak, A. (2010). Reducing costs of spot instances via checkpointing in the Amazon Elastic Compute Cloud. In *Proceedings of the 2010 IEEE 3rd International Conference on Cloud Computing*, pages 236–243.

[Yu and Magoulès, 2007] Yu, L. and Magoulès, F. (2007). A framework for dynamic deployment of scientific applications based on WSRF. In Cérin, C. and Li, K.-C., editors, *Advances in Grid and Pervasive Computing. Proceedings of the 2nd International Conference in Grid and Pervasive Computing (GPC 2007), Paris, France, May 2–4, 2007*, volume 4459 of *Lecture Notes in Computer Science (LNCS)*, pages 579–589. Springer-Verlag.

[Yu and Magoulès, 2008] Yu, L. and Magoulès, F. (2008). Towards dynamic integration, scheduling and rescheduling of scientific applications. *Journal of Algorithms and Computational Technologies*, 2(3):391–408.

[Yu and Magoulès, 2009] Yu, L. and Magoulès, F. (2009). Service scheduling and rescheduling in an applications integration framework. *Advances in Engineering Software*, 40(9):941–946.

[Zaharia et al., 2009] Zaharia, M., Borthakur, D., Sarma, J. S., Elmeleegy, K., Shenker, S., and Stoica, I. (2009). Job scheduling for multi-user MapReduce clusters. Technical Report UCB/EECS-2009-55, EECS Department, University of California, Berkeley.

[Zaharia et al., 2010] Zaharia, M., Borthakur, D., Sarma, J. S., Elmeleegy, K., Shenker, S., and Stoica, I. (2010). Delay scheduling: a simple technique for achieving locality and fairness in cluster scheduling. In *EuroSys'10: Proceedings of the ACM SIGOPS European Conference on Computer Systems 2010*, pages 265–278.

[Zaharia et al., 2008] Zaharia, M., Konwinski, A., Joseph, A., Katz, R., and Stoica, I. (2008). Improving MapReduce performance in heterogeneous environments. In Draves, R. and van Renesse, R., editors, *Proceedings of the 8th Symposium on Operating Systems Design and Implementation*, pages 29–42. USENIX Association.

[Zhang et al., 2009] Zhang, S., Han, J., Liu, Z., Wang, K., and Feng, S. (2009). Accelerating MapReduce with distributed memory cache. volume 0, pages 472–478, Los Alamitos, CA, USA. IEEE Computer Society.

[Zhimin and Vivek, 2005] Zhimin, C. and Vivek, N. (2005). Efficient computation of multiple group-by queries. In *SIGMOD'05: Proceedings of the 2005 ACM SIGMOD International Conference on Management of Data*, pages 263–274.

Index